Animal Partners and Parasites

Animal Partners and Parasites

PHILIP STREET

Illustrated by Nancy Lou Gahan

Taplinger Publishing Company | New York

First published in the United States in 1975 by
TAPLINGER PUBLISHING CO., INC.
New York, New York

Text copyright © 1975 by Philip Street
Illustrations copyright © 1975 by Taplinger Publishing Co., Inc.
and David & Charles (Holdings) Ltd
All rights reserved. Printed in the U.S.A.

Published simultaneously in the Dominion of Canada by
Burns & MacEachern, Ltd., Toronto

Library of Congress Catalog Card Number: 72-6630

ISBN 0-8008-0255-1

CONTENTS

ILLUSTRATIONS

INTRODUCTION

No animal species exists in complete isolation from all other species. There is always some interaction between species occupying the same environment. Sometimes they are competing for the same food supplies, and in many cases their relation is one of predator and prey. There are, however, many examples of more intimate relations between different species. These are of two main kinds—cooperation or partnership by commensalism or symbiosis, and antagonism through parasitism. The factor which distinguishes partnerships and parasitism on the one hand from more general relations on the other is that in the former the relation is between individuals, whereas in the latter it is between populations.

There are varying degrees of partnership and of parasitism. In many examples of partnership it is quite clear that both species involved gain mutual advantage from the relation, whereas in others the advantage to one partner may be quite clear but it is difficult to see what advantage the other gains. There are, too, examples in which it is difficult to see that either species benefits, though the association is so constant that it is difficult to imagine that it would persist if there was no advantage to either party.

Partnerships, however, may conveniently be divided into two main types. Commensalism is the term usually used to describe external partnerships, such, for example, as those between crabs and sea-anemones. Symbiosis embraces examples in which one partner lives within the body of the other; good examples are the ciliates which live in the gut of their ruminant hosts, and the flagellates which live in the intestine of wood-eating termites.

Parasites, too, can be conveniently divided into two main

groups—ectoparasites, which live on the surface of their hosts bodies; and endoparasites, which actually live within the host's body. Many ectoparasites are free-living creatures which only visit their hosts when they are in need of a meal.

There is a great variation, too, in the effects which parasites have upon their hosts. At one end of the scale there are mild parasites which cause little more than irritation to the host, and at the other parasites whose effects are so serious that they almost invariably cause the death of the host.

One interesting aspect of both partnerships and parasitism is that these are not evenly distributed through the animal kingdom. Some groups have many more examples of partnerships than others, and the same is true of parasitism.

Animal Partners and Parasites

CHAPTER ONE

BIRDS AS PARTNERS

The first suggestion that animals might enter into partnership for their mutual benefit was made by the Greek historian Herodotus some 2,500 years ago. He described how the courser, a bird now known as the Egyptian plover, provided a valuable service to the Nile crocodile by ridding it of the parasites with which it became infested. 'Living in the river', he wrote, 'the crocodile gets its mouth full of leeches, and when it comes out and opens its jaws to the westerly breezes, the courser goes in and gobbles up the leeches, which good office so pleases the crocodile that it does the courser no harm.'

At first sight this seems an extremely unlikely story, and in detail it may not be accurate. But Herodotus was correct in his assertion that a real partnership does exist between the crocodile and the bird. What is indisputably true is that the Egyptian plover does associate with the crocodile while it is sun-bathing on the river bank or on exposed sandbanks, feeding on the numerous parasites which infest the reptile's skin. And when a crocodile is basking, it usually holds its mouth wide open, so it is not unlikely that if a foraging plover sees a leech attached to the reptile's tongue, or a piece of meat wedged between its teeth, it may well hop into its mouth for such a tasty morsel. Naturalists in fact cannot make up their minds whether or not the plovers really do enter the crocodile's mouth. There might well be little danger in this apparently hazardous exploit, because when the crocodile is basking, it is usually well fed, and therefore perhaps unlikely to attack a bird which it might well pursue if it was hungry.

Besides ridding its partner of irritating parasites, the plover

1

acts as a sentinel. If danger approaches, the bird flies up and away and the crocodile, responding to the message, slides back into the water. It has also been recorded that if the crocodile is slow to respond to the vocal warning, the bird will return and peck it on the head.

A relative of the Egyptian plover, known as the thick-knee, also feeds on crocodiles' parasites, and nests near the spot where a crocodile has laid its eggs. Neither nest nor bird is ever molested by the crocodile. The immunity of the plover and the thick-knee are particularly interesting, because water birds form a considerable part of a crocodile's diet. Presumably in the nesting, as in the basking, situation, crocodiles are not interested in feeding.

Wart-hog with tick-birds

If there is some doubt as to the exact role played by the Egyptian plover, there is none so far as the African tick-birds or ox-peckers are concerned. These are smallish birds, about the size of the common starling, to which indeed they are quite

2

closely related. Their chosen partners are almost any of the larger game, such as buffaloes, rhinos, antelopes, wart-hogs and zebra—animals which are common in most parts of Africa—as well as domestic cattle. The partnership is always a very close one, the tick-bird virtually spending its life on its partner's back, except when it is disturbed and during the nesting season. Even when the host wades into the water to drink, the bird retains its position, unless it has chosen a wart-hog as its companion. Wart-hogs like to wallow in the mud, so the bird flies into a nearby tree, but returns as soon as the ablutions have been completed.

All the tick-bird's food is derived from its host, and consists of ticks buried in its skin and the numerous flies and other insects which land on its body. The partnership between the birds and their hosts must be of very long standing, because the birds have become structurally modified to be able to move with ease all over the host's body. Their claws are sharp and curved and their tails long and stiff like the woodpecker's, enabling them to cling to their host's flanks and to crawl under their bellies looking for ticks. Part of the bird's food is inevitably the partner's blood, with which the bodies of the ticks will be gorged. Nevertheless the hosts do benefit by having their tick population kept within bounds.

Even when they are not feeding, the birds use the backs of their hosts as platforms for sunbathing, and it is there, too, that they conduct their courtship displays and mating. At the breeding season the host even provides the nesting material, the nest being lined with hairs plucked from its skin.

Besides ridding their hosts of their ticks, the birds provide another valuable service. They act as sentinels to warn of the approach of danger—by uttering cries which become more insistent as the source of danger gets nearer. If the response is slow, the bird is said to peck its partner's head to reinforce the warning. The bond between rhinoceroses and tick-birds seems to be particularly strong, perhaps because the rhinoceros is very short-sighted, and is thus in greater need of the services of a sentinel than most other large mammals. Man is of course the

3

enemy most to be feared by the African game animals, and it is his approach they mostly need to be warned against. In this connection it is interesting that when tick-birds are perched on the backs of domestic cattle, they ignore the approach of man, neither flying away nor uttering warning cries.

There is some evidence that tick-birds have become physiologically adapted to a diet rich in ticks. They certainly have the reputation of being difficult to keep for long in captivity. It may well be that they need a certain amount of mammal blood which has been changed through the action of tick saliva.

Another bird which associates with African big game is the cattle egret, or buff-backed heron, to give it its correct name. It is a small species of heron, though considerably larger than the tick-bird. The partnership between it and the mammals is much less obligatory than the tick-bird's, but it does provide evidence of the probable way in which the latter partnership evolved.

Although the egrets may be seen perched on the backs of their partners, they are more often seen running around their legs as they walk slowly forward when they are grazing. The favourite food of the egrets is grasshoppers, and as these are disturbed by the feet of the grazing game and rise into the air, the birds grab them. Most of the grasshoppers are very well camouflaged, and the birds would have difficulty locating them as they lay concealed in the grass. A number of other birds—bee-eaters, for example—have also learned to take advantage of the insect food disturbed by the movements of grazing mammals, but in their case there is no partnership, the birds merely taking advantage of the conditions without contributing anything in return.

The cattle egrets do, so to speak, pay for their keep. While many of them may be chasing disturbed grasshoppers, others may be seen on the backs of the game pecking away at the ticks in their hides. They also act as sentinels, flying up from the backs of their partners at the first signs of danger.

The cattle egret is a much-travelled bird. Soon after the first white settlers arrived in Australia, it also appeared to provide the same service for their domestic cattle that it traditionally provided for the wild game animals of Africa. How it got there no one knows. Around 1880 some managed to cross the Atlantic to arrive in South America, whence they spread northwards to reach the eastern United States.

In the United States they have been able to change their habits to a certain extent. For example, they have formed associations with horses, sheep, goats, and pigs, as well as with domestic cattle. When these latter retire in the afternoon to chew the cud, the egrets come together in groups to forage for their own insects. This they do by leapfrogging over each other, each one disturbing insects for the individual over which it has jumped. In this way they all get a fair share of the insects which are disturbed.

In contrast to the cattle egret, the cow-bird is native to the United States. In earlier times it followed the enormous herds of bison as they undertook their spring and autumn migrations. When the bison became virtually extinct in the latter half of the last century, the cow-birds transferred their attentions to domestic cattle, following them to feed on the insects which they disturbed, just as they had previously followed the bison.

The cow-birds, however, had one feature not shared by any of the other birds which followed grazing animals. When the bison herds migrated south in the spring, the cow-birds could not afford to leave them to nest; or the birds would have lost the bison while they incubated and reared their fledgelings, and might never have established contact with them again. To solve this problem they did what the cuckoo does, and developed the habit of laying their eggs in the nests of other birds, which reared the eggs for them. When they transferred their attentions to domestic cattle, which do not of course migrate, they could well have reverted to the habit of incubating their own eggs, but they have not done this. So it is still left to other species to hatch out their eggs.

No bird partnership is more remarkable than that between the honey guide, a relative of the common cuckoo, and the ratel, an African species of badger. The ratel is a honey eater, breaking into the nests of wild bees, while the honey-guide feeds on bee grubs and beeswax. It cannot, however, break into the strongly built nests of the wild bees, which are always located in trees, but must rely upon the ratel to open the nests for it.

It does not, however, just wait for the ratel to discover a nest. Its method of procedure is to locate a nest, and then look for a ratel, calling all the time. The ratel seems to understand the message, and willingly follows where the honey-guide leads, eventually arriving at the tree where the bird has discovered the nest, which it proceeds to open. While it consumes the accumulated honey, the honey-guide feeds on the grubs and the wax.

Honey-guides, like cuckoos, indulge in what is known as brood parasitism, the females laying their eggs in the nests of other species. Sometimes the parent bird will destroy the other eggs in the nest when it lays its own, but more frequently it is left to the hatched honey-guide chick to get rid of its nest companions. However its potential nest companions are got rid of, the fact remains that the honey-guide chick is given no guidance as to its future role in life. It is not fed upon beeswax, nor is it given any help from its foster parents, and it certainly has no contact with its own parents, which might be assumed to train it for adulthood. So it must be purely by instinct that, as it becomes adult, it assumes the role of a honey-guide, able to lead the ratels to the nests of the wild bees. Perhaps the most remarkable feature of the honey-guide's behaviour is that it will lead humans to a wild bees' nest with the same efficient insistency as it leads the ratel.

In addition to the undoubted partnerships that exist there are a number of examples in which birds enter into associations which are much looser, though beneficial, at least to them. Small birds will often build their nests beneath the nests of larger birds of prey which would normally attack them. In the vicinity of their own nest, however, the predators seem somehow to suspend their aggressive instincts towards their neighbours. It is as though the bird of prey regards hunting as something which is done well away from its own home, not on its own

Honey-guide and ratel

doorstep, and other potential predators are unlikely to risk attacking the small birds when they have such powerful neighbours.

The osprey's nest is similarly used for protection by small birds. In this case there would be no direct risk to the small birds, since ospreys are fish eaters and do not normally attack birds; but the close proximity of the ospreys will serve to deter the normal predators of the small birds from approaching their nests. Birds like wrens and sparrows mostly take advantage of the protective canopy of the osprey's nest, but the night heron avails itself of the same protection. It is a fish-eating bird, and gains more than protection from the association, for when the parent ospreys are off the nest hunting, the heron visits it to eat any fragments of fish which may be left.

In Africa and south-east Asia members of the crow family known as drongos have a reputation for driving off birds of prey by attacking them viciously. A number of other birds take advantage of the drongo's disposition by nesting in the lower branches of the trees in which these crows have built their own nests. Here they are relatively safe from the attentions of birds which in normal circumstances would prey upon them.

In a similar quest for protection by proximity a number of birds build their nests near hornets' nests. In South Africa both the scarlet-breasted sunbird and the blue-breasted waxbill build within a few inches of the opening of a hornets' nest. Any other birds which come near the hornets' nest are attacked, but the birds in the neighbouring nest are never molested. Why these birds should gain this immunity is hard to see. They certainly gain by the association, because no potential predator would willingly incur the wrath of a nest of hornets, but it is difficult to postulate any advantage to the hornets.

It may seem incredible that an animal as large as a whale should be troubled by lice, but in fact its skin is pitted with the tiny depressions which these little creatures make and in which they lie to avoid the worst effects of the slipstream the giant mammal creates as it hurtles through the water. They are not real lice, which are insects, but tiny crustacean parasites (see Chapter 15).

The fact that these lice-covered whales should have birds to rid them of their unwelcome guests is perhaps even more incredible. As the antarctic winter approaches, the giant sperm whales migrate northwards to the warmer waters off the coasts of Africa and South America; and being mammals they must come to the surface periodically to breathe. Since each breath enables them to stay down for a considerable time, when they do surface, they also take quite a time to empty and refill their lungs.

It is then that they are located by the grey phalarope, a seabird which specialises in feeding on crustaceans, picking these up on the beach and from the surface waters as it skims over the waves. While it is thus employed, it often encounters a surfaced whale, and at once alights on its back to enjoy a feast of whale lice, presumably regarding the enormous back of the whale as a temporary beach.

So far all our examples have been of birds ridding other animals of their undesirable insect and other parasites. Our last example concerns birds which are rid of their own parasites through the activities of lizards. Many of the gulls which nest on the islands of the eastern Mediterranean become heavily infested with lice and other insect parasites during the nesting season. At this time the common wall lizard, which is very numerous on these islands, invades the nests and feasts on this abundance of parasites, to its own and the gulls' advantage. This is an interesting example, because normally gulls would kill and eat any small lizard they encountered; but the wall lizards go unmolested, as though the gulls recognised the advantage of having them around.

CHAPTER TWO

FISH PARTNERS

Many examples of fish entering into various degrees of partnership with other aquatic animals have been known for a long time, but the development of underwater research through the aqua-lung over the past few decades has helped to reveal how extensive such partnerships really are. Some of them bear a striking resemblance to certain of the bird partnerships we have been considering in the previous chapter. Fish, no less than any other animals, suffer from the attacks of skin parasites, and many of them have the equivalent of the mammals' tick-birds to help them deal with the problem.

Fish which behave like ox-peckers are known as cleaner fish. Thirty years ago almost nothing was known of their activities —indeed they were scarcely suspected—and it was not until 1949 that the systematic study of cleaner fish was begun by skin-diving biologists. They revealed a whole new range of fish activity. Today more than forty species of fish have been shown to take part in the cleaning routine.

Basically the cleaner fish goes all over the body of the fish which it is cleaning, picking off the fish lice and other skin parasites with its characteristic beak-like mouth and tweezer-like teeth. More than this, it will nibble away at patches of fungi and bacteria which may be infecting the skin, and if the fish has been injured, it will eat away any dead flesh and thus clean up the wound.

The advantage to the cleaner fish is obvious—it gets a good living from its cleaning activities. The host fish, of course, also benefits by having its irritating skin parasites removed. But does it show any awareness of the benefits which the cleaner

10

fish brings to it? The answer must be emphatically 'yes'. Many fish parasites settle on gills, and a fish being cleaned has often been observed to raise a gill cover to allow a cleaner fish to enter its gill chamber. When the cleaner fish emerges, the host fish raises the gill cover on the other side, and the cleaner fish swims to that side to clean the remainder of the gills.

The ultimate in recognition must be the tolerance shown by certain types of shark, which not only allow cleaner fish to enter their gill chambers to clear out the parasites, but will actually open their mouths so that they can enter and clear out their mouth parasites. Equally daring is the little Red Sea wrasse, whose body seldom exceeds 2in in length. One of these tiny fish has been seen busily picking parasites from the head of a 4ft moray eel, one of the most voracious of all fish. When its head had been cleaned, the eel then opened its mouth to allow its partner to enter, and kept it open until all the parasites had been removed; the tiny partner was then allowed to depart unmolested. Any other kind of small fish daring to approach a moray eel would have been snapped up immediately.

There is nothing haphazard about the cleaner fish meeting their customers. They usually have particular stations among the rocks where they can always be found, and it is to these stations that the other fish come when they feel in need of cleaning. Often there will be a queue in the vicinity of the station, and, as one satisfied customer swims away, so the next fish in the queue takes its place. Not only local fish come to the cleaners. Certain deep sea fish, notably the ocean sunfish, whose disc-shaped body may weigh a ton or more, make long journeys into inshore waters to avail themselves of the services of the tiny cleaner fish. Of course such giants need, and get, a whole shoal of cleaner fish working over their bodies.

It is now realised that the number of fish in any rocky locality may well depend upon the number of cleaner fish living there. As an experiment, skin divers removed all the cleaner fish from a reef. Within a short time the majority of the other fish had also vanished from the area, which was then virtually without any fish population. When the cleaner fish, which had been maintained in an aquarium, were returned to the reef, it was not long before the other fish, too, returned to their original habitats.

In recent years a scheme to dump derelict cars in the coastal waters off Florida was conceived, the idea being that, given additional shelter, fish would be encouraged to settle in the area. And this in fact they did, but not for the reason originally postulated. What happened was that the additional shelter encouraged the cleaner fish to move in and, once they were established, their customers also appeared, to swell the numbers of the resident population.

Interesting though these partnerships between cleaner fish and their customers may be, it is even more interesting to realise that there are exactly similar partnerships between certain shrimps and fish, the shrimps also being visited by fish needing cleaning. The best investigated example of these shrimp cleaners is the Pederson shrimp, *Periclimenes pedersoni,* which always lives in association with the sea-anemone *Bartholomea annulata,* either attached to the anemone's body or living in the same rock cavity with it.

Here is an account of the activities of the Pederson shrimp by the late Conrad Limbaugh, who did so much to further our knowledge of cleaner fish and cleaner shrimps:

When a fish approaches, the shrimp will whip its long antennae and sway its body back and forth. If the fish is interested it will swim directly to the shrimp and stop an inch or so away. The fish usually presents its head of a gill cover for cleaning, but if it is bothered by something out of the ordinary, such as an injury near its tail, it presents itself tail first. The shrimp swims or crawls forward, climbs aboard and walks rapidly over the fish, checking irregularities, tugging at parasites with its claws and cleaning injured areas. The fish remains almost motionless during this inspection and allows the shrimp to make minor incisions in order to get at subcutaneous parasites. As the shrimp approaches the gill covers, the fish opens each one in turn and allows the shrimp to enter and forage among the gills. The shrimp is even permitted to enter and leave the fish's mouth cavity. Local fishes quickly learn the location of these shrimps. They line or crowd round for their turn and often wait to be cleaned when the shrimp has retired into the hole beside the anemone.

The Pederson shrimp is solitary, but another cleaner, the California cleaner shrimp, *Hippolysmata californica,* is gregarious. During the daytime it hides away in rock crevices, but after nightfall it roams the sea bed in search of customers. Almost any fish it encounters is thoroughly gone over until all its external parasites have been removed.

Much less rigid than the cleaner partnerships are those associations between defenceless fish and more powerful animals which can offer them protection from their potential predators. In inshore waters rocks provide a good deal of protection, but for fish which live in the open seas there are no such natural hiding places. In these open waters, though, large jellyfish abound, and with their formidable stinging tentacles they are usually given a wide berth by even large fish. Many small fish have learned to take shelter beneath the umbrellas of these jellyfish, swimming among the tentacles which hang down from them. So far as we know the little fish sheltering under the umbrellas are just as vulnerable to stings from the tentacles as other fish which habitually avoid coming into contact with them. Only skilful swimming keeps them out of trouble.

Much more dangerous than the true jellyfish is a distant relative of theirs, the Portuguese man-o'-war, whose body consists of an air filled bladder which floats on the surface of the sea and from which long streamer-like tentacles hang down into the water. The sting batteries on these tentacles are particularly powerful, capable of stinging a man very severely. To any smaller animal, including quite large fish, they are fatal. Yet there is one small fish, appropriately known as the man-o'-war fish, which habitually lives among these tentacles. Whether these fish have acquired some kind of immunity to the sting poison is at present uncertain. They are known to feed on bits

Clown anemone fish with anemone

of the tentacles which they bite off, but at the same time they are also known to be killed by the man-o'-war and eaten by it. More research is needed before we have the complete story of this particular partnership.

14

A much more involved partnership exists between certain kinds of fish and sea-anemones, which are sedentary relatives of the jellyfish and have similar stinging tentacles. It has been known for a long time that certain small fish generally called damsel fish or clown fish could live among the tentacles of the large anemones of tropical coral reefs, where both are to be found. It was at one time thought that these fish had a natural immunity to the poison injected by the anemones' sting cells, but recent research has suggested that the relation is much more complicated than this.

Any damsel fish coming in contact with the tentacles of any anemone other than the one with which it normally lives can be killed and devoured by it. What appears to happen is that each fish has to establish its own immunity relation with its chosen anemone. When a fish first encounters its chosen partner, it does so with extreme caution, getting gradually closer and closer to the anemone until it makes brief contact with one or more of the tentacles. This causes the tentacles to react violently, while at the same time the fish jerks away from the first tentative grip. A series of further contacts follow, and with each one the reaction of both the fish and the anemone becomes less vigorous, until eventually the fish is able to swim among the anemone's tentacles and touch them with seemingly complete immunity.

What apparently happens during this acclimatisation process is that the mucus on the skin of the fish acts as an inhibitor of the anemone's sting cells, so that eventually contact with the particular fish invokes no response at all. But any other fish, with a skin mucus to which it has not been acclimatised, can be killed by its sting cells.

The advantage to the fish is obvious. It is able to live among the deadly sea-anemone tentacles where no other fish would dare to attack it. But what advantage does the anemone gain? This seems to be twofold. The fish probably keeps the anemone clean by ridding it of all debris, and it may also act as a decoy, luring other fish within range of the deadly tentacles.

One of the best known associations is that between sharks

and manta rays and the small pilot fish which habitually swim with them. In earlier times it was believed that these pilot fish guided their larger partners to their prey, but it is now known that sharks are quite able to seek out their prey for themselves. The partnership seems to be nothing more than a desire on the part of the pilot fish to swim within the cover provided by a much larger object. Thor Heyerdahl, in his account of the Kon-Tiki expedition, describes how the pilot fish which had been swimming with sharks which were harpooned by the members of the expedition attached themselves to the raft after they had lost their original companions, so that finally the raft was being escorted by a considerable shoal of these fish.

The pilot fish gain considerable advantages from the association. Besides finding immunity from attack by their potential enemies, who would scarcely come within range of the sharks' formidable jaws, they also get a share of any meal which the shark catches. Sharks are messy feeders, many pieces dropping down into the water when they are dealing with their prey, and these are snapped up by the pilot fish.

A number of traditional beliefs surround these pilot fish. In remote times they were regarded as sacred, and it was believed that they would accompany ships, dispersing only when the ship was approaching land. This gave a warning to the sailors of land ahead, especially useful if the weather was bad, for they would be unable to see it for themselves. There was also a belief that pilot fish would swim around a shark approaching a baited hook in order to persuade it not to do so. In fact on more than one occasion pilot fish have been seen apparently guiding their shark towards bait on a hook. When the shark has been hooked, the pilot fish have been observed swimming off, presumably to look for another partner.

There are many example in the sea of hitch-hikers, small fish which attach themselves to larger fish for the ride—and the advantages which result from such an association. In the waters of the Pacific there is a fish known as the emperor snapper which is about 2ft in length. Often it has as an attached travelling companion a much smaller fish known as the spotted grouper, which fixes itself to its larger companion by means of a suctorial mouth. So long as the snapper is actively swimming, the grouper maintains its hold, but when the former stops to feed, it detaches itself and swims around its companion picking up scraps of its food, or even catching small fish which may swim by. But the partnership is not one-sided, for the smaller fish acts also as a cleaner, picking off any parasites which may develop on its partner's skin.

The best known of these fish hitch-hikers is the sucker fish, remora. This fish has, on the top of its head, a large oval sucker which is corrugated like the sole of a tennis shoe, and represents a very modified dorsal fin. With this sucker it adheres to the under surface of a shark or a giant manta ray. The sucker is extremely efficient, and can only be detached by the application of considerable force. Like the spotted grouper, the remora is able to share in its host's meals and also to catch small fish for itself. In return for the mobility and safety it gains from the partnership, it keeps the skin of its host free from external parasites. It is also known to enter the gill openings and even the mouth of its partner for this purpose.

Altogether there are about ten different species of these sucker fish, each species choosing a different species of fish as its host. Some even attach themselves to marine turtles. And they vary in size from a few inches to about 3ft. They have been known for a very long time, the ancient Greeks believing they could slow down or even stop a ship by attaching themselves to its hull.

For a long time, maybe for thousands of years, remoras have been used by fishermen in various parts of the world to catch fish for them. The method used is to attach a rope harness on a long line to the sucker fish and then throw it overboard. It will soon attach itself to a passing fish, when the line can be drawn in. The attachment of the sucker is so firm that the captured fish has no chance to escape, and can be easily hauled on board. Besides fish, marine turtles are also captured by this means. Sometimes the turtle is too big and powerful to be hauled in to the boat's side, in which case one of the crew dives overboard to attach a powerful rope to it, so that those remaining on board can haul it in.

Sharing house is a common form of partnership between marine animals. In the Indo-Pacific area, for example, small gobies take shelter in the burrows of snapping shrimps. These burrows, excavated in sand, need almost constant attention to prevent them from becoming filled in, and the shrimps spend much of their time on this task of digging. The gobies spend most of their time just above the entrance to the burrow waiting for suitable food organisms to come along, but at the first signs of danger they dive into the burrow, and this also serves to warn the shrimps, which will never emerge from their burrows unless their attendant gobies are outside on guard. Each burrow usually provides shelter for a pair of gobies.

Another species of goby, the arrow-goby, *Cleviandia ios,* which lives along the coast of North America, enters into more complicated home-sharing partnerships with various burrowing invertebrates. The Echiuroida is a small phylum of unsegmented marine worms containing only about seventy different species. One of the best known of these, and one of the largest, is *Urechis caupo,* which lives in U-shaped burrows in the same areas as the arrow-goby. And it goes in for lodgers in a big way. At the bottom of the burrow there is often a pea crab, and somewhere along its length there may well be also a scale worm and a small species of clam. And there are the gobies. As many as two dozen have been found sheltering in a single burrow. The crabs, scale worms and clams all benefit from the respiratory current of water which the echiuroid is continually drawing through the burrow, because from it they filter out food particles as well as receiving adequate supplies of oxygen. For their food requirements the gobies make sorties from the mouth of the burrow into the surrounding water.

The really interesting thing about this multiple partnership is that the worm clearly tolerates its various partners, for if any other kind of animal attempts to enter the burrow, it will try to keep it out. The partners, too, almost certainly seek out the burrow of this particular worm and not those of other kinds, although in the mud in which the echiuroid lives, there are plenty of other burrowing worms. To find the burrow, they probably rely upon chemical sense similar to taste or smell, and the worm also recognises them in the same way.

We know that many marine invertebrates have such a chemical sense. Barnacles settling on the bottoms of ships can be a nuisance, and a great deal of research has been devoted to finding methods of discouraging the barnacle larvae from settling. Out of this research has come the fact that barnacle larvae will settle more readily on surfaces where other barnacles are already growing than on surfaces where there are none. It seems clear that the resident barnacles must give out some chemical substance into the water which serves to attract the larvae.

Recent research has revealed another example in which some kind of chemical sense guides an animal to the burrow of its partner. A crab, *Pinnixa chaetopterana,* normally shares the burrow of a marine worm, *Chaetopterus pergementaceus.* In experiments the crabs were put into a stream of sea water which had previously been passed over a number of the worms. The effect was dramatic. The crabs were roused to considerable activity, attempting to move towards the area of greatest concentration of whatever substance was given off by the worms. Under natural conditions such reactions would lead the crabs to the burrows of the worms.

Besides being attracted to echiuroid burrows, gobies are attracted to the burrows of various species of burrowing shrimps. One of these, *Callianassa affinis,* provides shelter and a home for the blind goby, *Typhlogobius californiensis.* When it is young, it is normally coloured and has good eyesight, so that it can swim actively about in search of its food. But as it grows up so it pales to a flesh colour, its eyes become covered with an opaque layer of skin, and it becomes blind. Before this occurs, it has forsaken the free life and taken up residence in a shrimp burrow. Usually a male and a female share the burrow.

In this particular example the burrow is more than a home or a shelter—it is the fish's complete territory, which it never leaves. If any rival male tries to enter the burrow, the resident male drives it away with great vigour. The shrimps feed on detritus (minute particles of organic matter) which they sift out from the surrounding sand. The gobies live on any small animals which find their way into the burrow, and upon broken pieces of seaweed which also fall into it. They may well therefore play a part in keeping the burrow clean.

On coral reefs in the warmer seas there is a kind of sea-urchin known as the hat-pin sea-urchin because it has immensely long, sharply pointed spines which guarantee it against attack from almost any other marine creature. Two kinds of fish—the shrimp-fish and the cling-fish—take advantage of this immunity by living among these spines. Although the two kinds of fish are not at all closely related, they have both become much modified to fit them for their unusual mode of life. Their bodies are long and thin, whereas the bodies of their close relatives are of normal fish shape, and they swim vertically among the spines with their heads pointing downwards. Here they are quite safe from the attacks of potential predators. They are rendered inconspicuous by their bold black and white pattern, which breaks up the outline of the body.

Not content with taking shelter in other animals' burrows, there are certain other species of fish which actually hide away within the bodies of other animals. Sponges are particularly prone to these lodgers, the structure of their bodies making them particularly useful for this kind of sheltering. The whole body of a sponge is riddled with canals, some of which take in a respiratory and feeding current of water while others act as the channels by which the current leaves after it has supplied the sponge's needs.

These canals make ideal places in which fish can hide to escape from the attentions of predators. Underwater exploration, especially around the West Indian islands, has revealed a whole new world of fishes. Early on, it was discovered that when a particular kind of finger-shaped sponge was touched, almost invariably one or more tiny fish would dart out of the openings of the canals. These were mainly gobies, blennies and damsel fish. By plugging the vents with cotton wool and then taking the sponges to the surface it was possible to make a closer study of the fish. Many of those which were found to be quite common in these sponge retreats belonged to a species which previously had been thought to be quite rare, while others had never been seen before, and were therefore new to science. Collectively, we now know, sponges harbour a vast fish population, and in the larger specimens an incredible number of individual fish may be sheltering.

The sponge-blenny, *Paraclinus marmoratus,* not only shelters in sponges but also breeds there. The female produces eggs which stick to the sides of the canals. These egg masses are guarded until they hatch by the male fish, which fan them with their fins to keep them aerated.

Shrimps belonging to a number of different species also shelter in sponges. One of the largest sponges found in Caribbean waters is the loggerhead sponge, which can be as large as a barrel. A moderate-sized specimen examined was found to contain no fewer than 16,000 shrimps.

One family of small fish, the Carapidae, known as pearl-fish, have become adapted structurally and functionally to living within the bodies of other marine animals, including bivalve molluscs, sea-urchins, starfish and sea-squirts, but their favourite partners are various kinds of sea-cucumbers. These are echino-

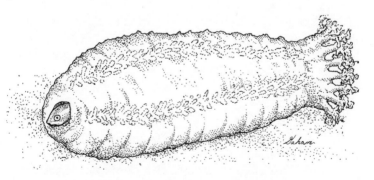

Pearl-fish living in sea-cucumber

21

derms, and are thus related to the sea-urchins and the starfish. They have cylindrical bodies, and move slowly about on the sea bed. A number of thin-walled branched tubes open into the cloaca or hind gut. These tubes are the creature's respiratory organs, which are rhythmically filled and emptied with sea water by a current which is taken in and passed out through the large cloacal opening.

The best known of these pearl-fish is the Mediterranean species, *Carapus acus,* which resides either in the cloaca of its partner or in the respiratory tubes, sometimes even rupturing these to get right into the body cavity. Any damage caused in this way to the sea-cucumber is not really important, because sea-cucumbers have a great capacity for regenerating lost or damaged parts of their internal anatomy. They can even lose virtually the whole of their internal organs and regenerate them in quite a short time without any apparent inconvenience.

The pearl-fish pass freely in and out of their living home through the cloacal opening, usually emerging to feed on small crustacea and other animals. But when they are within the sea-cucumber, they have no respiratory difficulties because they can take advantage of their partner's respiratory current. Sometimes they will supplement their diet by feeding on some of the partner's internal organs. When returning to shelter, adult fish usually enter the cloacal opening tail first with a kind of cork-screw motion, but young fish will often enter head first. Sea-cucumbers are in fact very good partners, providing a high degree of protection for the pearl-fish. The mucus secreted by their skin is poisonous, and for this reason very few animals will attempt to molest or eat them.

Pearl-fish have become modified in a number of ways. Most fish have scales, which makes them smooth to the touch if they are stroked from head to tail, but rough and resistant if stroked in the opposite direction. The pearl-fish, however, have lost their scales, and this makes it easy for them to slide in backwards through the partner's cloacal opening quite easily. They have also lost their hind or pelvic fins, which would get in the way. Their anus, too, opens towards the front, so that in

22

order to void faecal waste, the fish has only to expose the front part of its body through the cloacal opening.

Adult pearl-fish can and do live free lives without entering into partnership with sea-cucumbers, although those which do so probably fare better than those which do not; but there is evidence that during a certain period of their life history a partnership is necessary. Pearl-fish experience two larval stages —first, living in plankton, and secondly, finding a suitable partner before changing into a young adult.

Our last example is a partnership between two animals, not for their own advantage but for that of their offspring. The animals concerned are a European freshwater fish known as the bitterling *(Rhodeus amarus)* and freshwater mussels. As the breeding season approaches, the egg duct of the female bitterling begins to elongate until it is about 2in in length. At the same time she pairs up with a mate, and they both approach a mussel. Now mussels draw in a continuous current of water through a tube known as the inhalant siphon, this current providing them with both their oxygen and their food requirements. The female bitterling lays her eggs near the entrance to this siphon, so that they are drawn into the body of the mussel. The male follows and sheds his seminal fluid, which is also drawn into the siphon. Inside the shell of the mussel the bitterling eggs are fertilised, and there they remain until they hatch.

At the same time as the bitterling are producing their fertilised eggs, the mussel is also reproducing, so that the mussel eggs hatch at the same time as those of the fish. The mussel larvae now attach themselves to the young fish, and when these leave the shelter of the mussel shell, they take the young mussels with them. The mussel larvae bury themselves beneath the skin of the young fish, and here they remain until they have developed into young mussels. Then they bore their way out and sink to the river bed, where they can settle and grow into adult mussels well away from their parent colony.

CHAPTER THREE

PARTNERSHIPS INVOLVING CRABS

Crabs are another group in which partnerships with other animals are quite a common habit. The best known of all these crab partnerships are those between hermit crabs and a variety of other species. Whereas most animals only enter into partnership with one other species, the hermit crab is often found with two partners of quite different species.

Despite its name the hermit crab is more closely related to the lobsters than to the typical crabs, having the well developed muscular abdomen which lobsters use to propel themselves swiftly backwards through the water. In crabs the abdomen has become reduced to a small remnant which is carried permanently turned forward beneath the carapace.

Whereas the abdomen of the lobster is heavily armoured with a thick shell like the rest of the body, in the hermit crab the abdominal shell has been lost, leaving the abdomen unprotected. It is also permanently twisted so that it fits easily into the coils of the snail shell in which the crab lives for protection. The last pair of abdominal appendages have become modified to form sickle-shaped hooks which are used to anchor the abdomen to the central pillar of the adopted shell. On the thorax only the first two pairs of walking legs are well developed, and used for walking, the other two pairs being much reduced in size and kept permanently within the shell.

If danger threatens the front part of the body, which is well armoured and usually projects beyond the mouth of the snail shell, it is withdrawn with a jerk, leaving only a claw visible, blocking the entrance to the shell. This is always the right claw, which is much larger than the left one tucked in behind it.

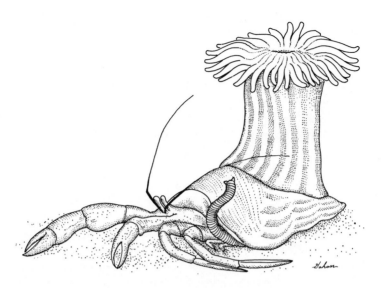

Hermit crab with *Calliactis parasitica* anemone and *Nereis furcata* worm

One of the largest of the hermit crabs is *Eupagurus bern-hardus,* with a body up to 6in long. When it is full grown, only a large whelk shell is large enough to accommodate it. Almost every specimen will carry an anemone belonging to the species *Calliactis parasitica* firmly fixed to the top of the adopted whelk shell, and many of them will have as a second guest one of the ragworms, *Nereis furcata,* which lives inside the shell alongside the crab's body. Although ragworms are equipped with power-ful jaws, their bodies are soft and vulnerable; *Nereis* thus gains valuable protection by living in association with the hermit crab. Feeding is also made easy for it. Whenever the crab is feeding, the worm thrusts its head out of the shell and shares its meal.

Philip Henry Gosse, in his famous book *The Aquarium* (1856), was the first to describe this association between the hermit crab and the worm. He has been describing the association between the hermit and its attendant anemone, and goes on :

But I find that this association is not the only one that exists here. While I was feeding one of my Soldiers, by giving him a fragment of cooked meat, which he, having seized with one claw, had transferred to the foot-jaws (maxillipeds), and was munching, I saw protude from between the body of the Crab and the Whelk-shell the head of a beautiful worm, *Nereis billineata* (now called *furcata*), which rapidly glided out round the Crab's left cheek, and, passing between the upper and lower foot-jaws, seized the morsel of food, and, retreating, forcibly dragged it from the Crab's very mouth. I beheld this with amazement, admiring that, though the Crab sought to recover his hold, he manifested not the least sign of anger at the actions of the Worm. I had afterwards many opportunities of seeing this scene enacted over again; indeed; on every occasion that I fed the Crab and watched it eating, the Worm appeared after a few moments, aware, probably by the vibrations of its huge fellow-tenant's body, that feeding was going on, and not, I think, by any sense of smell. The mode and place of the Worm's appearance were the same in every case, and it invariably glided to the Crab's mouth between the two left foot-jaws.

There can be no question about the very considerable advantage the worm gains from this partnership. But does the crab gain anything? It has always been suggested that the worm helps by keeping the inside of the shell clean, and it certainly would consume any materials that did gain entrance. By rhythmical movements of its body the worm draws into the shell a respiratory current of water for its own use, and this might well also be of similar benefit to the crab.

But the main contribution of the ragworms is providing empty whelk shells for the hermit crabs to live in. Whelks are preyed upon by dogfish, which bite off their heads and as much of their bodies as protrude beyond the entrance to their shell as they crawl along the sea bed. That part of the body remaining within the coils of the shell, however, is beyond the reach of the fish, and is left behind. It will not be long before this store of fresh food is found by a wandering *Nereis furcata,* which will make its home in the well stocked whelk shell. As they grow, so hermit crabs are always searching for larger shells to replace the ones they are living in, and long before the worm has managed to consume the remains of the whelk, it is likely that an almost full grown *Eupagurus* searching for a final home will also have found the shell, and will settle itself in with its future partner already in residence.

The partnership between *Eupagurus* and *Nereis* is not universal, only a minority of the crabs having worm partners; but with the anemone *Calliactis parasitica* the partnership is much more constant, the crab seldom being found without at least one specimen of the anemone firmly fixed to the top of the whelk shell and sometimes as many as three. *Calliactis* is a widespread species, and is found in partnership with a variety of different species of hermit crabs throughout the world.

What exactly each partner gains from this particular association is not easy to say, but from its widespread occurrence there must be some mutual advantage. It has always been said that the crab must gain protection because fish will not risk attacking it for fear of coming into contact with the deadly sting cells on the anemone's tentacles, and this may well be true for many species of fish. On the other hand, certain fish which habitually prey upon hermit crabs and are skilled at dragging them out of their homes apparently take no notice of the anemones, and will still attack a crab even when there are three of them on the whelk shell. The anemone may also provide a certain amount of camouflage, making it more difficult for a potential enemy of the hermit crab to detect the whelk shell when it has an anemone sitting on the top of it.

The possible advantage to the anemone is equally difficult to establish with any certainty. It was thought at one time that when the crab was feeding, the anemone bent forward until its tentacles were in the vicinity of the crab's mouth, and that they

27

were able to pick up particles of food which escaped the crab's own jaws. But we now know that when the crab is feeding, and often at other times as well, the anemone bends backwards and not forwards, so that its tentacles are sweeping the sea bed behind the shell. Here they may well be able to pick up plenty of food material stirred up by the movements of the crab.

The fact that there is evidence to show that the crabs and anemones will go out of their way to achieve their associations suggests that they recognise the advantages even if we cannot define them with certainty. If a crab has to change its shell for a larger one, it has on many occasions been seen to detach the anemone from the old shell and place it carefully on the new one. On other occasions the anemone has transferred itself to the new shell.

A much more permanent association is one between the hermit crab *Eupagurus prideauxi* and the cloak anemone *Adamsia palliata*. This crab is a much smaller species of hermit, seldom exceeding 2in long. While it is still quite small, and therefore occupying a small shell, it usually acquires its partner. But the anemone, instead of settling on the back of the shell, settles on its under surface just behind its opening, and therefore also behind the crab's mouth. Its basal disc then gradually grows round the shell until its two edges meet over its back, so that the shell is now completely enclosed by the anemone.

From now on the two animals increase in size together, and as the crab becomes too large to be completely enclosed by the shell, so the basal disc of the anemone grows forward beyond the mouth of the shell to increase its effective capacity. The crab thus comes to lie partly within the original shell and partly within the tube formed by the extended basal disc of the anemone. In this way the crab escapes the periodic necessity of moving to a larger shell. This is always a hazardous procedure, for when a hermit crab removes its abdomen from one shell, it is very vulnerable until it has succeeded in tucking it away into the new one.

Certain tropical hermit crabs live in the depths of the ocean, some of them more than 2,000 fathoms below the surface, where it is completely dark. But the crabs carry round on the backs of their borrowed shells anemones which produce phosphorescent light. Many crabs living in the depths are completely blind, but these hermit crabs have well developed eyes. Whether the pale light produced by their attendant anemones gives sufficient illumination for them to detect and pursue their prey is not certain, but it may well serve to attract the crabs to each other at the breeding season.

The third kind of animal living in partnership with certain hermit crabs is the sponge. Unlike the two earlier species of *Eupagurus,* which are often found in seashore rock pools, *Eupagurus cuanensis* is only found in shallow offshore waters. Early in its life this hermit crab usually chooses a shell which already has a small specimen of the sponge *Ficulina ficus* growing on it. If it happens to occupy a shell without a sponge, it goes in search of one, and when it has found one, it detaches it and plants it on its own shell.

As both animals continue to grow, the sponge gradually extends beyond the shell until it has grown right round the crab, leaving a sufficient opening through which the crab's head, claws and two pairs of walking legs can still emerge. When the crabs are hauled up in a dredge, they look just like round orange-red sponges. Often only after they have been placed on deck and put their legs through the single tiny opening and begin moving about are they recognised for what they are.

Although the sponge can offer the crab no active protection against attack from its enemies, being completely devoid of weapons of any kind, it is probably of considerable benefit as camouflage for it. Sponges are little sought after for food, only certain species of sea-slugs showing any interest in them, and they could certainly do no harm to the hermit crab.

The fact that partnership with sponges is beneficial to crabs seems likely in view of the number of different and not very closely related species of crabs which wear sponges on their carapaces. One of the best known of these is *Dromia vulgaris,* commonly called the sponge crab. It is a Mediterranean species which may extend as far north as the southern Cornish coast, and is similar in general appearance to the edible crab, though much smaller, seldom having a carapace wider than 3in. Its last pair of walking legs, however, are turned permanently upwards over the back of the carapace, and each ends in a small pair of pincers.

Like the hermit crab it too chooses *Ficulina ficus* as its partner. When it is quite young, the crab searches for a small specimen of the sponge and, when it has found it, carefully places it on its carapace. This is well supplied with setae, giving it a hairy appearance, and these, together with the last pair of legs, serve to hold the sponge securely in place. As the crab grows, so also does the sponge, until it forms a yellow cap fitting right over the carapace and concealing the crab beneath. When at rest with its legs and claws tucked away beneath it, the crab is quite invisible to any enemy viewing it from above.

Besides *Dromia* there are many different kinds of sponge crabs in various parts of the world, and recent studies have revealed how some of them manage to cover themselves with sponges. They are able to cut out from a large sponge a piece which almost exactly fits their carapace, using their claws as scissors. Having achieved the right shape and size, they then hollow out the under surface so that it fits the curved back of the carapace. A captive crab's sponge was removed and placed on the floor of the aquarium tank, but it was able to replace it without much difficulty. When the sponge was placed in another tank with a number of other sponges and the crab subsequently transferred to this tank, it was always able to recognise and replace its own sponge. In another experiment in which the sponges were removed from a number of sponge crabs and replaced by sheets of paper the crabs even cut out pieces of paper, of the right shape, and tried to anchor these to their carapaces. Clearly these crabs' partnership with sponges is part of their way of life.

Our last example of crabs using sponges is somewhat different from the others in that the sponges used are only one of a number of organisms which are attached to the crabs to provide camouflage. Spider crabs are easily distinguishable from all other crabs by their very long thin legs and relatively small bodies (which are usually more or less triangular in shape). Their shells are provided with hooked hairs which make the attachment of animals and plants easy.

Spider crabs show considerable ability in dressing up, always attaching to themselves plants and animals which blend with their surroundings and make them as inconspicuous as possible. If a specimen should move permanently into different surroundings, the old camouflage will be removed and a new set of plants and animals attached. Once, when a specimen wearing green seaweeds in addition to sponges and other sedentary animals was put into an aquarium tank where the weeds were all red, it spent many hours laboriously removing the green weeds one by one and replacing them with red ones. To enable a spider crab to do this dressing up its claws are able to reach every part of its carapace, in contrast to other crabs, and this is a point worth remembering when a spider crab is handled.

What degree of protection against its enemies a hermit crab derives from the anemones which live on its borrowed shell is uncertain, as already mentioned, but there is no doubt as to the value of sea-anemones to certain crabs inhabiting the Indian and Pacific oceans. There are two species of these grenadier crabs, as they are sometimes called, which are quite widely separated geographically but have remarkably similar habits. *Lybid tessellata* lives in the coastal waters around the Seychelles Islands, while *Polydectes cupulifera* is found off Hawaii.

These crabs carry anemones around with them, one in each claw, choosing any one of a number of different species. The protection they provide for the crabs is not just passive, ie discouraging their enemies from approaching. If one should appear to be coming in to attack, the crab will stretch out its claws holding the anemones towards it so that it faces two rings of anemone tentacles, which effectively bar its way. There is some evidence to suggest that the crab also uses its partners as food gatherers. If one of them captures a small fish on its tentacles

31

and does not push it through its own mouth fairly quickly, it may well be stolen by the crab.

This particular partnership must be of very long standing, since the crab's behaviour has been modified in adaptation to it. Normally all crabs use their claws for picking up and conveying food to their mouths, but the grenadier crabs have transferred this function to their first pair of walking legs. It is said that if a crab comes across a larger anemone than one of those it is carrying, it will set its own specimen down and pick up the larger one. Another species of crab living around the coasts of Japan does not actually hold its anemone partners but carries around instead two bivalve mollusc shells on each of which an anemone is living.

Pea crabs are the smallest of the world's true crabs. Many different species are known from many parts of the world, and all of them live for protection with a variety of other marine animals. One of the best known species is the European pea crab, *Pinnotheres pisum,* which has been recognised since early times as a partner of various molluscs. The males grow to a maximum breadth of $\frac{1}{4}$in, while the females are somewhat larger, attaining a maximum breadth of $\frac{1}{3}$in. For protection they take up residence within the shells of living mussels, oysters and cockles, the majority apparently preferring mussels.

The way in which the pea crabs behaved within their mollusc shells was investigated by the ingenious device of cutting windows in the shells of molluscs which had been adopted by pea crabs. These molluscs feed on diatoms and other microscopic members of the plankton. A respiratory and feeding current of water is drawn into the mollusc shell and passed through the perforated gills, which hang like curtains down either side of the animal's body. The suspended plankton organisms are trapped on the gill surfaces, which are covered with mucus. This mucus, laden with food, is conveyed towards the mouth as a continuous moving string, the mucus current and the respiratory current both being kept moving by the continuous beating of myriads of minute cilia projecting from the surface of the gills. The pea crab, lying handily placed between the gill curtains, is in the right position to feed on the mucus current whenever it feels hungry.

Full grown females sometimes become too large to be able to leave the mollusc shell even if they wanted to, but this is really no hardship to them, even at the breeding season. At this time the males leave their adopted homes temporarily to visit and mate with the females. The fertilised eggs are then shed among the gill curtains of the mollusc and carried out of the shell in the outgoing respiratory current. For a time the eggs, and the series of larvae which are typically produced during the development of a crab, float in the plankton near the surface of the sea. When the last larvae finally turn into small crabs, these sink to the sea bed and seek their partners. The width of the carapace at this time is only about 1mm. At first they do not seek members of the species which will provide them with their final homes, but hide away in the shells of *Spisula solida,* one of the bivalve molluscs known as trough shells. These are burrowing species like cockles, and live just below the surface of sand or gravel on the lowest parts of the shore and beneath the shallow offshore waters beyond. After some time, when they will of course have grown considerably, the little pea crabs leave their first hosts and seek the hosts within whose shells they will spend the remainder of their lives.

The female of another tiny crab also finds itself a prisoner in the home when it reaches full size. This is the gall crab, *Hapalocarcinus,* which is found only in tropical seas, where it forms a partnership with coral. Coral consists of huge colonies of animals looking very much like tiny sea-anemones, to which in fact they are closely related. Unlike anemones, however, each one builds a hard cup of limestone around its base. Because the members of the colonies live very close together these cups join up to form great masses of limestone. Each generation of coral animals forms its cups on top of those left by earlier generations, and in this way massive coral reefs are built up.

The young female gall crab chooses for its future home a small space surrounded by young coral animals. As they grow, the corals adjust the shape of their cups so as to form a cavity in which the crab can live in complete safety, making it large enough for the crab when it has eventually grown to full size. Small openings are left through which the imprisoned female crab can draw in a current of water for feeding and respiration, gall crabs being plankton feeders. The males live free on the coral reef, and, as they are much smaller than the females, they can pass through the opening into the cavities to visit and mate with them at the breeding season.

Animals which remove skin parasites from their partners are not confined to birds and fish. Among the most interesting of the animals found in the Galapagos Islands are the large marine iguanas, large lizards which feed on seaweed but spend a good deal of their time sunning themselves on the rocks. Also living on these rocks are the red rock crabs, and these can be seen walking all over the iguanas as they sleep. They are looking for the ticks which are quite common but not easily detected on the lizards' skins. When one is spotted, the crab can be seen tugging away at it with its claws until at last it is pulled out.

CHAPTER FOUR

ANT GUESTS AND PARTNERS

However common partnerships may be among other groups of animals, none can rival the ants in the number and variety of the animals which enter into some kind of association with them and often live in their nests. These latter are usually referred to as the ants' guests. Many of these are parasites, preying upon the ants or their young without giving anything in return, but many others are true partners.

The best known of all the ants' partners are the aphids—the greenfly and the blackfly—from which the ants get the sweet honeydew which they like so much. The way in which ants 'milk' the aphids to persuade them to produce drops of the coveted liquid was first described by a Swiss authority on ants, Pierre Huber, in his book *The Natural History of Ants* (1820). He refers to the aphids as pucerons, the name by which they were known at that time.

I observed a branch of thistle covered with brown ants and pucerons and noticed that the latter regulated the time when they discharged their excreta. I remarked that it very rarely passed at the natural period and that the pucerons, stationed at some distance from the ants, scattered it afar off, by a movement somewhat resembling the kicking or wincing of a horse. How happened it, then, that the ants, wandering upon the branches, displayed bellies remarkable for their size, and evidently filled with some kind of liquid. This is what I learned by closely watching a single ant whose movements I am about to describe. I saw it at first pass, without stopping, some pucerons, which it did not, however, disturb. It shortly

35

after stationed itself near one of the smallest and appeared to caress it by touching the extremity of its body alternately with its feelers, with an extremely rapid movement. I saw with much surprise, the fluid proceed from the body of the puceron and the ant take it in its mouth. Its feelers were afterwards directed to a much larger puceron than the first, which, on being caressed after the same manner, discharged the nourishing fluid in greater quantity, which the ant immediately swallowed. It then passed to a third which it caressed, like the preceding ones, by giving it gentle blows with the feelers on the posterior extremity, of the body; the liquid was ejected at the same moment and the ant lapped it up. It then proceeded to a fourth; this, probably already exhausted, resisted its action. The ant, who in all probability knew it had nothing to hope for by remaining there, quitted it for a fifth from which it obtained its expected supply. It now returned perfectly contented to its nest.

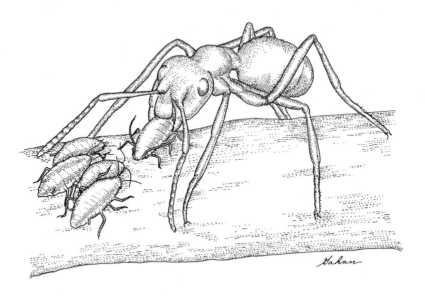

Ant with aphids

The aphids, or plant lice as they are also called, are a very successful and widespread group of insects found in most parts of the world, and almost everywhere they occur there are ants which milk them for their honeydew. The question inevitably arises, why do these tiny insects exude this clear sweet liquid through their anuses instead of normal faeces? The answer is to be found in their feeding habits. Their mouthparts have become modified for piercing and sucking, and they use them to penetrate the soft parts of plants and imbibe the plant juices. Now these are very rich in sugars, but have only minute quantities of the other food materials which are just as essential to the health and well-being of the aphids. So in order to obtain sufficient protein and other necessary food materials the aphids have to suck up relatively enormous quantities of plant juice, and this means that they swallow vastly more sugar than they need. It is this surplus sugar, together with the water in which it is absorbed, that is passed out through the anus as honeydew. Ants in their turn show a great desire for sugar, which they have learned to get by milking the aphids. There is some evidence to suggest that aphids which are milked regularly by their attendant ants retain their honeydew until they are stroked, whereas those without regular attention will exude the sticky liquid over the surface of the plant as and when it accumulates inside their bodies.

While many ant species are content to go out among the plants growing near their nests to milk the aphids living on them, quite a number take active steps to protect their adopted aphids. The simplest way of doing this, used by many species is to carry fine soil up the plant stems and construct, with the aid of their saliva, tiny earth sheds into which the aphids can retreat and thus hide from their potential enemies. If there are insufficient aphids on the plants near their nests, ants will travel further afield and bring back to the nearer plants such aphids as they can find to reinforce their own populations. In some cases the ants will collect their aphids at the end of the day and bring them into their nests for protection overnight, returning them to their food plants in the morning.

37

Not all aphids live above ground. Some species live below ground and feed on plant roots, and these are milked in the same way as those species living above ground on plant shoots. These also may be collected from further afield and settled on the plant roots that run through the ants' nest.

Sir John Lubbock, a famous Victorian naturalist whose book *Ants, Bees and Wasps* became a natural history classic, was the first to show that small black objects found in the nests of certain ant species were in fact aphis eggs. We now know that the common yellow ant collects aphis eggs from the food plants on which they have been laid and carries them to its nest, where they will be protected through the winter. In the following spring they are taken out of the nest by the ants and placed on their correct food plant, so that when they hatch they will provide a handy source of honeydew for the ant colony.

A particularly interesting example of egg collecting comes from America, where certain kinds of ants have as partners aphids which live on corn roots. Their eggs are collected towards the end of the summer and taken to the ants' nests for protection during the winter. When they hatch out during the following spring, they come too soon for the corn roots, which are not sufficiently developed to provide them with food. Consequently the ants take them to the roots of other plants for the time being, transferring them to the corn when it has grown sufficiently to support them.

Aphids have two closely related insects which also live on plant juices—the scale-insects and the mealy bugs. These also produce honeydew, and many of them are milked by ants. Perhaps the most interesting of these are the pineapple mealy bugs, which live on the pineapples grown in Hawaii. Unless these are milked regularly by their attendant ants, they become drowned in their own exudations of honeydew. So the Hawaiian pineapple growers do not have to spray the mealy bugs in order to destroy them. The more effective way of dealing with what is indeed a serious pest is to eliminate the colonies of the attendant ants. A pineapple field which is clear of ants will soon lose its mealy bugs, but one which has no mealy bugs will soon acquire plenty of them, because any ants present will bring them from neighbouring fields in which they are abundant.

Although ants are avid collectors of honeydew, they have not, in contrast to the bees, learned to store the sweet liquid so that it can be available for future use in feeding themselves or their offspring. There are no honeydew combs. When ants return to the nest after a collecting trip, they can only pass over some of the honeydew to those of their fellows who have remained behind to perform the many tasks that have to be done.

In some parts of the world, notably South Africa, Australia and the southern states of North America, there are species of ants in which certain individuals have become modified as living stores of honeydew. These specialised workers, which are known as honey-pot ants or repletes, are living storage tanks. They are able to swallow immense quantities of honeydew from which other members of the nest are able to satisfy their needs. Repletes are much too laden with honeydew to be able to move about. They live in certain parts of the nest, where they usually rest suspended from the roof.

Ants it seems will do anything in return for a supply of sweet liquid, and aphids and their relatives are by no means the only animals capable of producing sweet secretions. Certain beetles are also capable of producing a sweet liquid similar to honey-dew. One kind of rove beetle, *Lomechusa strumosa,* makes a comfortable living for itself out of its ability to produce such a secretion. It is often found in the nests of slave-making ants. These ants steal eggs from the nests of other species, and when these hatch out, force the transplanted ants to work for them. Presumably because they do not, therefore, rely to the same extent as other colonies on producing their own workers, these slave-making ants seem quite content for the rove beetles to feed upon their own eggs and larvae. In return the beetles provide plentiful supplies of the coveted sweet liquid.

Many of the guests which live in ants' nests provide nothing for the ants, but since they do no serious harm to them, other than stealing some of their food, they are not serious parasites.

39

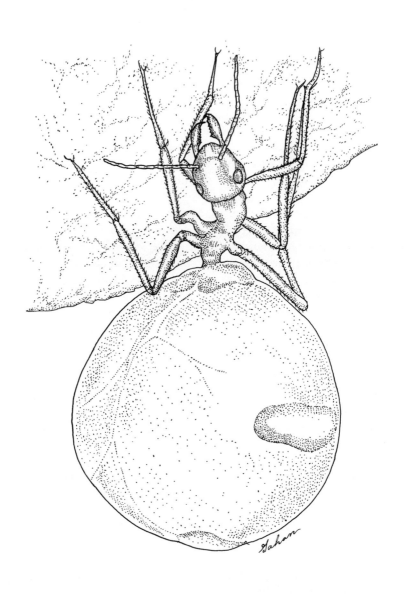

Honey ant

Among these are various species of silverfish, tiny insects which we sometimes find in the larder. When ants which have been out collecting honeydew return to the nest and pass some of it to their fellow ants that have remained behind, the silverfish will appear on the scene to share in the exchange.

The larva of a particular species of fly found in Texas lives in ants' nests, where it attaches itself by means of a sucker to the neck of an ant larva. When an adult ant comes along to pass a drop of honeydew to the young ant, the fly larva protrudes its mouth and steals some of the drop. It is said that if the fly larva is hungry and there appears to be no meal in the offing, it will bite the ant larva's skin, thus causing it to wriggle, which is taken as a sign by the adult ants that it is hungry, and will bring one of them running to give it a sustaining drop of sweetness.

Another feeding partner of ants is the mite, which, like the silverfish, obtains its food from its partner without giving anything in return, but at the same time cannot really be classified as a parasite because it does no harm to its ant partner. Some ant species have but one mite guest, which usually clings beneath its mouth. When it is hungry, it scrapes the ant's mouth with the first of its four pairs of legs, and this stimulates the ant to produce a drop of honeydew, which the mite proceeds to lap up. Some ants carry a number of mites, but these are always arranged symmetrically so that the ant is not overloaded on one side, and some of these may beg food from other ants that may be alongside their own ant. Some mites have even been observed to jump from their own ant on to another which happens to be passing, this presumably being a means of avoiding overcrowding. One kind of mite does not take food regurgitated from the ant's mouth; instead it eats the small particles of food the ant brushes off its own body with the small comb-like structures on the last segments of its legs, which it uses to keep its body clean.

Perhaps the most remarkable association involving ants is that between certain species, especially the two common red ants, *Myrmica scabrinoides* and *Myrmica laevonoides,* and the large blue butterfly, *Maculinea arion.* By the end of the nine-teenth century the life histories of all the British butterflies were known except for that of the large blue. It was known that its caterpillars lived on wild thyme plants until the third moult, feeding on the flowers, but that at some time after this moult they disappeared, no one knew where. Caterpillars generally undergo four moults before pupating to form a chrysalis, but no large blue chrysalides or caterpillars after the fourth moult had ever been found. The third moult caterpillars just disappeared, yet the following summer a new generation of adults appeared as if by magic.

Attempts were made to breed the caterpillars in captivity, as had been successfully achieved for every other species of British butterfly. Up to the third moult they thrived on their natural diet of wild thyme flowers, but after this moult they became increasingly restless, wandering aimlessly about and ceasing to feed, until finally they died.

The story of how the mystery of the large blue butterfly was finally solved is one of those fascinating chapters of natural history. In the summer of 1915, F. W. Frowhawk and Dr Chapman, two butterfly experts on a visit to Cornwall, turned their attention to collecting large blue caterpillars from wild thyme. One of these they found growing on the top of an ants' nest. For some reason they pulled it up and saw, lying in the exposed nest, a butterfly chrysalis. Now they were quite capable of recognising and identifying the chrysalis of every British butterfly, except of course that of the large blue, which no one had ever seen, and this chrysalis which they had unearthed was one they had never seen before. It must, they felt, be that of the large blue.

To prove this they initiated a series of experiments, in which they constructed an experimental ant hill with thyme plants and large blue caterpillars. As a result of their patient observations they were able to piece together the whole story of the blue butterfly.

After the third moult the caterpillar leaves the thyme plant on which it has been feeding and begins to wander aimlessly about. In time it will probably encounter an ant, which will immediately show great interest in it. By this time it will have developed a gland on the dorsal surface of the seventh abdominal segment, which is the tenth behind the head, and this gland, sometimes known as the honey gland, secretes a sweet liquid similar to the honeydew secreted by the aphids.

The ant which discovers it will go immediately to this gland and caress it with its antennae. This will cause the gland to secrete a drop of its sweet liquid, which is eagerly lapped up by the ant. This process will be repeated a number of times until the caterpillar suddenly changes its behaviour. It hunches up its body, and the ant picks it up by the scruff of its neck and carries it back to its nest.

Once installed in the nest, the caterpillar is treated as an honoured guest, the ants feeding it with their own grubs in return for the coveted sweet liquid it produces. After a number of weeks, when it has grown to full size, it ceases to feed and goes into hibernation within the safety of the ants' nest. In the following spring it wakes up again, feeds for a time, and then pupates to form a chrysalis, from which it emerges a few weeks later as an adult butterfly. In most butterflies, as soon as the adult emerges from the chrysalis, case blood is pumped into the folded wings, causing them to expand. But in the large blue butterfly this process is delayed to enable it to leave the ants' nest and climb to the top of a plant stem, where there is room for its wings to expand and dry out.

All subsequent attempts to rear large blue caterpillars after the third moult on any other food other than ant grubs have failed, suggesting that the grubs must contain some food materials which are essential to the later stages of the large blue's development. In this partnership the butterfly certainly gains more than the ant, but it is another example of the sacrifice which ants are prepared to make in return for drops of sweet liquid.

Although it is the large blue which enters into such intimate association with ants, a number of other blue butterflies also possess the sweet gland, and are attended by ants, though none of them are taken to the ants' nest. Nevertheless they do probably benefit from the association, since their potential enemies will be likely to hesitate to attack them when there are ants around. The caterpillars of the chalk-hill blue, *Lysandra coridon,* and the silver-studded blue, *Plebejus argus,* are sometimes farmed by ants. They carry the caterpillars in their mouths and place them on suitable food plants near their nests.

CHAPTER FIVE

MICROSCOPIC PARTNERS

Our last examples concern partnerships in which one of the parties is a unicellular organism living within the body of its much larger partner. In some cases there is a difference of opinion as to whether the smaller partner is an animal or a plant, because at this level of organisation it is not always easy to differentiate between the two.

One of the most interesting examples is the partnership between some termites and certain flagellate Protozoa. It is unfortunate that termites are often commonly called white ants, because they are quite unrelated to the true ants. The members of the insect order Hymenoptera, to which the ants belong, are among the most highly developed of all insects, whereas the order Isoptera, which consists of the termites, is a relatively primitive insect group.

There are two groups of termites—a more advanced type which live in the ground and build the spectacular termite hills which are a feature in many of the warmer parts of the world, and more primitive types which live in wood. The latter are the species which cause wooden buildings and other structures to collapse without warning because they have reduced them to paper-like shells by eating away all the wood beneath the surface.

The major constituent of wood is cellulose, with small amounts of sugars and other carbohydrates, and animals generally are not provided with a cellulose-digesting enzyme. Wood-boring beetles and other insects are unable to utilise the cellulose, only the small proportions of sugars and other carbohydrates. There are some species of Protozoa, however, which are able to digest cellulose. Among these are some flagellates which live in the hind gut of the wood-eating termites. They break down the cellulose into simpler carbohydrates which the termites are then able to digest. All worker and soldier termites must have flagellate colonies, but the king and queen termites do not possess them. They are given a special diet prepared by the workers.

These symbiotic flagellates have become so specialised for living in the hind gut of termites, where the oxygen concentration is extremely low, that they are incapable of living outside the termite body, where the oxygen concentration is normal. Termites which are experimentally exposed to a raised oxygen concentration soon die of starvation because their gut flagellates are killed, and they are therefore no longer able to digest the wood which is their only food. The termites are able to withstand higher temperatures than their partners, but if these are killed by being exposed to too high a temperature the termites again die of starvation.

When young termites are hatched, they will, of course, lack a flagellate colony; but they soon remedy this by eating some of the faeces of older members of the colony, which will naturally be rich in the Protozoa. When young queens and males leave the nest for their nuptial flight, they carry some of the infected faeces from the colony with them so that they will be able to infect the first members of their subsequent brood.

There is a very similar relation between cattle and other ruminants and certain species of ciliate Protozoa which live in the rumen or paunch. These ciliates cannot live in an atmosphere containing oxygen, but they thrive in the paunch, where the atmosphere is almost pure methane. They are able to break down the cellulose which forms the walls of cells and use it to satisfy their own energy requirements. Their growth is extremely rapid, the total life span being about 24hr. As a result they are also dying rapidly, and the vast quantities of dead ciliates provide a considerable quantity of food for the ruminants. It has been calculated that they obtain up to 20 per cent of their protein requirements by digesting the ciliates.

46

Our last examples of internal symbiosis concern certain single-celled algae in partnership with larger animals. Some of these, known as zoochlorellae, are green, but there are also yellowish or brown ones known as zooxanthellae. The latter are only found in partnership with marine animals, whereas the former are mainly found in association with freshwater animals.

One of the most interesting examples of these partnerships, and one which has become a classic, is that between a tiny flatworm known as *Convoluta roscoffensis* and certain green algae. *Convoluta* lives on and just beneath the surface of the sand on the shores of north-west France. Each individual is only a few millimetres long, but there can be such numbers that the whole shore takes on a greenish tinge. When the tide comes in, they burrow between the sand grains just beneath the surface for protection against the movement of the water.

When the young flatworms are first hatched out, they are colourless and completely devoid of any algal partners. But they

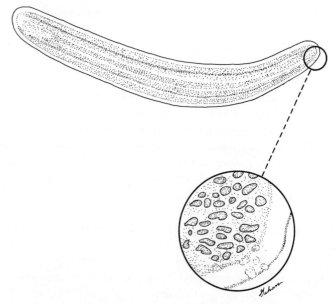

Flatworm, *Convoluta roscoffensis* with algae, *Chlorella*

47

do feed on the green single-celled algae which live and multiply attached to the sand grains. Many of these will be digested in the normal way as food, but some of them escape this fate and succeed in invading the flatworms' tissues, where they grow and multiply, and it is these which form their partners and give them their green coloration. For a time the partnership works perfectly, the algae growing and multiplying and the flatworms consuming the surplus to supplement their diet.

Later, however, the partnership seems to go drastically wrong. After the flatworm has matured and laid its eggs, the rate at which it consumes its algal partners is increased, so that these are gradually eliminated; this process begins at the tail end, which becomes white, and then the green colour disappears progressively forwards until the whole worm becomes white. At the same time or even earlier the worm's gut degenerates from the hind end, so that it is no longer able to eat food and pass it through the gut. Having at the same time lost its functional digestive system and its algal partners, there is nothing left for it but to die. Death is probably not caused solely by starvation. The symbiotic algae, besides providing it with food, also absorb its excretory products for their own use, so that, when they are eliminated, these products accumulate and poison the worm.

A close relative of *Convoluta roscoffensis* is the worm *Convoluta convoluta,* which does not lose its gut nor consume its symbiotic algae, so that the partnership persists throughout its life. Although the worm can eat other food, evidence suggests that it cannot survive indefinitely without its green partners, and that without them it fails to reach maturity and therefore cannot breed.

The part which zooxanthellae play in the lives of reef-building corals has been the subject of a great deal of research during the past 10 years or so. For some time it had been assumed that the brown algae which occurred regularly in the tissues of these corals played some part in their nutrition, and that they would be particularly valuable in removing their hosts' excretory products and utilising them themselves. Recent work by Japanese marine biologists, however, has shown that in some species at least the algae do not live inside but between the cells of the coral polyps, and that they excrete a substance which can be absorbed by the polyp cells to be used as food. First indications are that this substance is some kind of carbohydrate. It has also been shown that the photosynthetic activities of the algal partners of the coral polyps increase their ability to deposit carbonate, which is of course the material of which the coral reefs are made.

Among the most spectacular of all invertebrate animals are the giant clams, bivalve molluscs related to the mussels, oysters and cockles but immensely larger than their relatives. They form a special family of bivalves, the Tridacnidae, and they are found only on coral reefs in the warmer parts of the Indian and Pacific oceans. The largest member of the family, *Tridacna derasa,* can achieve a length of up to 6ft. In addition to their enormous size these giant clams have another claim to fame. They rely for food largely upon their partnership with enormous numbers of zooxanthellae. In adaptation to this partnership their bodies have undergone considerable modification. Whereas other bivalves have inhalent and exhalent siphons for the intake

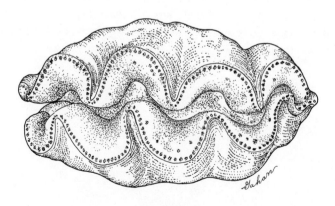

Tridacna clam

49

and exit of a continuous current of water at the hind end of the body, in the tridacnids these siphons have become rotated so that they face upwards, and the tissue surrounding them has become much enlarged.

It is this tissue that the zooxanthellae are found in their thousands, absorbing the waste products of the tridacnids' metabolism for their own use and producing enormous quantities of oxygen which make the bivalves' metabolism more efficient. As the algae multiply, so some of them are used as food by the molluscs, but never to such an extent as to threaten their collective existence. The molluscs are also able, like other bivalves, to filter microscopic plankton organisms from the current of water which they draw in through their inhalent siphons.

The tissue surrounding the siphons is coloured, usually peacock blue or moss green, giving the appearance of rich velvet when seen from above as the animals lie in shallow water. This dark colouring serves a strictly practical purpose, preventing the intense rays of the sun from penetrating the deeper tissues, where they might well cause damage.

Scattered among the siphonal tissue are numerous lens-like structures probably derived from the simple visual organs possessed by certain active bivalves, notably the scallops, which are used by them as eyes. In the tridacnids they serve to concentrate the light on dense clusters of zooxanthellae, presumably making them more efficient by providing them with increased amounts of light.

CHAPTER SIX

PROTOZOAN PARASITES

For the remainder of this book we shall be considering associations between pairs of animals in which one member of the pair is harmful to the other, and therefore a parasite. Some of these parasites are only mildly harmful, whereas other are pathogenic, that is they are responsible for some kind of disease in their hosts. The proportion of parasitic members varies very considerably from phylum to phylum. In some phyla parasites are rare, whereas in most others, although they may contain many important parasitic species, these are far outnumbered by the free-living non-parasitic members. In only one phylum, the Platyhelminthes, are there more parasitic than free-living species. Between them animal parasites are responsible for a considerable proportion of the most serious diseases both of man and his domestic animals.

The first group to be considered is the phylum Protozoa, whose members can be regarded either as unicellular or noncellular. Whichever view is taken, the fact remains that an individual protozoan has a structure and size which are both very similar to those of any single cell of a multicellular or metazoan animal; and just as there is considerable variation in size between the smallest and the largest cells in the body of a metazoan, so among the protozoans there are giant species and pygmy species. All, however, are microscopic in size.

The phylum Protozoa comprises four subphyla. In the following summary the principal characteristics and the most important human parasites are given.

The most important members of the subphylum Sarcodina from the human point of view are the amoebae, which are

51

familiar to anyone who has ever studied biology at school. There are many different species, the majority of which are free-living in the soil or in water. The minority of parasitic species, with one exception, live in the large intestine of the host, and several are parasites of man. Of these only *Entamoeba histolytica* is seriously pathogenic, being the organism responsible for amoebic dysentery. The remainder do little or no harm, but since they apparently do no good to their hosts either, we must regard them as mild parasites rather than as partners. Amoebae can move only slowly by a flowing movement of the cell protoplasm.

The second subphylum, the Mastigophora, comprises those Protozoa known as flagellates, whose most characteristic feature is the possession of one or two, occasionally more, whip-like outgrowths of the protoplasm known as flagella. These flagella are capable of making powerful lashing movements which propel the creatures along with considerable rapidity through water or any liquid medium in which they may live. The parasitic flagellates fall naturally into two groups—the intestinal flagellates which live in the gut of the host, and the haemoflagellates which live in the blood. By far the most important flagellate parasites of man are the haemoflagellates, the trypanosomes and their relatives, which are highly pathogenic. Because they have complicated life histories involving insect intermediate hosts we shall deal with them in Chapter 11.

In contrast to the other three subphyla, in which only a minority of species are parasitic, all the members of the subphylum Sporozoa are parasitic, with complicated life histories. From the human point of view the most important Sporozoa are those species responsible for causing malaria, but because like the trypanosomes their life cycles involve insect intermediate hosts they will also be considered in Chapter 11.

The final subphylum, the Ciliophora, contains the most highly organised of the Protozoa. They are active creatures, capable of moving quite rapidly through a fluid medium by the rhythmical beating of large numbers of minute cilia which cover virtually the whole of their surface. Cilia are similar in

52

structure and function to flagella but they are very much smaller. The vast majority of Ciliophora are free-swimming creatures. Only a small minority are parasites, but others live as partners in the intestines of many different kinds of animal.

Amoebae live generally by flowing round and ingesting micro-organisms smaller than themselves. In the parasitic amoebae these food organisms are almost invariably bacteria, and the place in the body where bacteria are most likely to be found is the large intestine, where considerable bacterial destruction of the indigestible part of the host's food is taking place. This explains why all but one of the human parasitic amoebae live in the large intestine. Few parasitic amoebae are found in the large intestines of carnivores, where only small numbers of bacteria are found.

The life histories of the intestinal amoebae are all very similar. The adults living in the intestine divide simply as soon as they have reached full size. They do not spend the whole of their life cycles in the intestine. At some stage they become transformed into cysts, in which each amoeba becomes surrounded by a resistant cyst wall. These cysts now have no power to retain themselves within the large intestine, and are passed out of the host's body with the faeces. While inside the cyst, the nucleus divides, usually twice, so that the cyst contains four nuclei. These cysts are clearly the means by which the amoebae leave their original host in an endeavour to colonise other hosts. If unprotected amoebae were passed in the faeces, they would be unable to survive for more than a very short time, whereas within the protective cyst walls they are able to survive for much longer, thus increasing their chances of being picked up and ingested by another host. It is probably, too, that without the protection of a cyst wall they would be killed and digested in the new host's stomach without getting a chance of reaching the large intestine.

When an encysted amoeba is ingested, the cyst wall is not dissolved away until the amoeba has reached the small intestine, and the enzymes found there are incapable of killing it. The four nuclei now divide again to give a total of eight, and this is followed by the division of the cytoplasm, so that the final result is eight small amoebae which travel down to the large intestine, where they grow to full size.

As already mentioned, the only serious pathogenic amoeba in man is *Entamoeba histolytica,* which causes a serious disease called amoebic dysentery. Most intestinal amoebae merely feed on such bacteria and other suitable similarly-sized particles which they can find, but *Entamoeba histolytica* is able to produce powerful protein-digesting enzymes capable of dissolving and feeding on the cells in the wall of the large intestine. The specific name *histolytica* means 'tissue dissolving'. As the cells are killed and digested, cavities are formed which soon develop into ulcers, and from these come a steady stream of blood and mucus, which of course pass out of the body with the faeces.

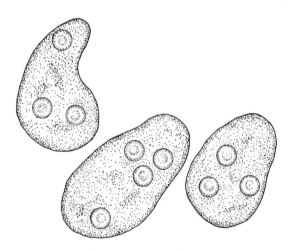

Trophozoites of *Entamoeba histolytica*

The passing of liquid faeces containing large amounts of blood is a sure sign of dysentery. Sometimes the amoebae penetrate so deeply into the intestine wall that they are able to enter the main blood circulation and so are carried round the body. When this happens, they may settle either in the liver or the lungs where they also cause abscesses.

The most puzzling feature of infection by *Entamoeba histolytica* is that only a minority of infected people in fact suffer from dysentery. In the majority, although there may be plenty of the amoebae flourishing in the large intestine, these cause only minor symptoms or none at all, and they do not eat their way into the intestine wall, remaining content with the food provided by the bacteria found there, like other intestinal amoebae which are not pathogenic. The reason for this dual behaviour is not known for certain.

What is known, however, is that there are two forms of *Entamoeba histolytica,* a larger and a smaller, and it is believed that the latter kind is not pathogenic. There is some evidence, too, that the larger form may itself consist of two races, only one of which may be pathogenic. If this is so, there are three different forms of *Entamoeba,* only one of which is capable of causing dysentery.

It was at one time thought that *Entamoeba histolytica* existed mainly in the tropics, but it is now known to be almost worldwide in its distribution. Dysentery itself is in the main found only in tropical climates; whether this is owing to the virtual absence of the pathogenic strain from temperate areas or to some other factor is not known. Certainly *Entamoeba* is spread more readily in tropical than in temperate areas. As we shall see repeatedly in the following chapters, the continued prevalence of all kinds of parasitic diseases in many tropical areas of the world can usually be put down to the extremely primitive conditions of sanitation and hygiene prevailing there, and this is almost certainly the reason why amoebic dysentery is so widespread in these areas. When faeces are deposited almost anywhere, the cysts can easily get into the water supply, which in general will of course be natural and not piped, and thence into the human beings. In many places, too, where both water and fertiliser are in short supply, sewage may be collected in some sort of container and then spread on ground which is being cultivated. In this way any vegetables grown in this ground will very quickly become contaminated with cysts.

Although *Entamoeba histolytica* is the only undoubtedly pathogenic amoeba in man, four other species with widespread distribution are also frequently found in the large intestine. Only one of these, *Dientamoeba fragilis,* has ever been suspected of being pathogenic, the other three apparently living and thriving in the large intestine on bacteria and small particles of debris without causing any inconvenience. The most common of these harmless intestinal amoebae is *Entamoeba coli.* Its distribution is virtually worldwide, it having been estimated that 30–50 per cent of the total world population may act as hosts to this species. Two other species quite frequently found in the human large intestine are *Endolimax nana* and *Iodamoeba bütschlii.*

Of all the various amoebae which live parasitically in man and other animals, only one does not live in the large intestine. This is *Entamoeba gingivalis,* which lives in the mouth not only of man but also of dogs, cats, monkeys, horses and probably many other kinds of animal. It feeds on bacteria and food particles which are generally available in the mouth, especially between the teeth, and between the teeth and the gums. It is a very common parasite, but there is no evidence that it is ever pathogenic. The fact that it is often present in unusually large numbers in cases of pyorrhoea is no proof of its participation in the disease. The probable explanation is that any kind of mouth infection probably results in an abundance of bacteria on which the amoebae can feed.

The parasitic members of the flagellates or Mastigophora are most conveniently divided according to the parts of the host's body in which they normally live rather than according to their zoological classification. The haemoflagellates live in the blood of their hosts (see Chapter 11). Here we will consider only the less harmful intestinal flagellates, some of which in fact live in parts of the body other than the gut—for example, the vagina—but never in the blood.

The shape of the intestinal flagellates is ovoid, with a single flagellum projecting from the hind end and usually three, though sometimes four or five, projecting as a bunch at the anterior end. Along one side of the body runs an undulating membrane, a kind of frill raised from the general cell surface. The passage of waves along this membrane supplements the beating of the flagella in propelling the flagellates through a liquid medium.

Most of the important flagellates intestinal parasites belong to the genus Trichomonas, known generally as the trichomonads.

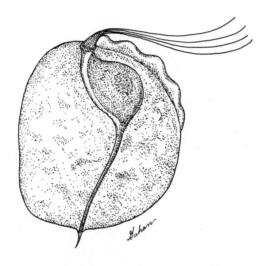

Trichomonas vaginalis (flagellate)

One of the most important of these is *Trichomonas vaginalis,* which is a very common human parasite. Unlike the majority of the so-called intestinal flagellates, it is not an inhabitant of the intestine but lives in the vagina, where it is very common, its incidence being variously estimated at 20–40 per cent. It is not a seriously pathogenic parasite, but it does produce congestion and irritation in the vagina; and there is some evidence that the death rate at childbirth is higher among mothers infected with *Trichomonas vaginalis* than among those free from it.

Another important human trichomonad 'intestinal' parasite which does not in fact inhabit the intestine is *Trichomonas tenax*. This lives in the mouth, and although it is more common in mouths where there is pyorrhoea or dental caries, there is no firm evidence that it is pathogenic. The probable explanation of its greater frequency in cases of mouth infection is that infected mouths provide more bacteria for it to feed upon, like *Entamoeba gingivalis,* as we have already seen.

When we come to the true intestinal flagellates, we find that the species most commonly found in the human gut is *Trichomonas hominis.* There is not much evidence that it is pathogenic, though some authorities believe that its presence may be associated with persistent diarrhoea.

The only other important intestinal flagellate affecting man is *Giardia intestinalis,* which differs from the other intestinal flagellates in two respects. It lives in the duodenum and the upper parts of the small intestine, and it produces cysts, which of course pass out with the faeces and constitute the infective stage of the parasite. The other flagellates are transmitted from one host to another by normal individuals which pass out with the faeces, but these cannot survive outside the body for as long as the cysts of *Giardia.*

In an infection *Giardia* occurs in very large numbers, and since these numerous flagellates adhere to the surface of the intestinal cells, they interfere with these cells' normal activities. They prevent them from carrying out their functions of absorbing fats, in particular, and also other products of digestion. This inhibition of normal absorption has two effects. It reduces the total intake of food into the body, so tending to cause malnutrition and deficiencies of the fat-soluble vitamins A and D. If fat absorption is reduced, so is the intake of the vitamins which it contains. The second effect is the result of the reduced fat absorption—persistent diarrhoea with faeces rich in the unabsorbed fat. Some idea of the astronomical numbers of flagellates there must be in the small intestine during a moderate infection can be gained from the fact that a single stool may contain more than 300 million cysts, and that in one severe infection the estimated number of cysts in a single stool was an almost unbelievable 14,000 million!

A number of flagellates that are parasites of animals also deserve mention. The first of these is a trichomonad, *Trichomonas gallinae,* a parasite of pigeons and other birds, including chickens and turkeys. It lives in the anterior part of the digestive system, from the mouth to the crop, whence it can invade various parts of the head, lungs and liver, with fatal results. The method used by pigeons to feed their nestlings favours the transmission of the parasite. The parent birds feed their young on partially digested regurgitated food which is known as pigeon's milk, and if the parent bird is infected with *Trichomonas gallinae,* it is of course transmitted to the offspring along with this milk.

Another trichomonad, *Trichomonas foetus,* is an important parasite of cattle, with almost worldwide distribution. It lives in the reproductive tract, and is transmitted by bulls to cows during mating. It is thus a venereal disease. The effects on the bull are minimal, but in the cow the mucus membrane of the vagina is attacked and the uterus invaded, causing abortions and stillbirths. Fortunately infected cows eventually acquire immunity and freedom from the disease, but bulls remain infected, able to transmit the disease throughout their lives.

Perhaps, however, the most important flagellate parasite of domestic animals is *Histomonas meleagridis,* which attacks all kinds of domestic birds but is particularly virulent in turkeys, in which it causes an infectious enterohepatitis commonly known as blackhead. In such infections the parasite is found both in the gut and in the liver. Turkeys are much more liable to pick up the infection from chickens than from their own kind, which makes it inadvisable to have chickens and turkeys running together.

Whereas the majority of the members of the other three subphyla are free-living, the parasitic species being in the minority, all the members of the subphylum Sporozoa are parasites. The most characteristic feature of the Sporozoa is a complicated life history during which there is an alternation of generations and intermediate as well as final hosts. Transmission in many cases is by special resistant forms called spores. By far the most important members of the subphylum are the various species of *Plasmodium,* which are responsible for malaria. Consideration of these species is deferred until Chapter 12, where the complicated life history of the parasite is described. Since the life histories of all the Sporozoa are essentially similar there would be no point in repeating the account here.

The subphylum contains a large number of types parasitising a great variety of different animals, invertebrates as well as vertebrates, but apart from *Plasmodium* and certain other human blood parasites they are of little medical or economic importance.

The members of the last subphylum, the Ciliophora, represent the most highly organised of all the Protozoa. With their covering of cilia they are active creatures, the free-living and the parasitic members alike requiring a liquid or a semi-liquid medium in which to live. The majority of the parasitic species seem to be harmless, and some of them, as we have seen in Chapter 5, are important partners of ruminants, in whose stomachs they digest the cellulose walls of plant cells to release the enclosed cell contents for digestion by the mammals.

Only one species, *Balantidium coli,* is pathogenic in man. It has the distinction of being the largest of all the human protozoan parasites. There is the same mystery surrounding the effects of *Balantidium* infections as there is with infection by *Entamoeba histolytica.* In many cases even a large scale infection of the large intestine produces no adverse symptoms, but in others the parasites acquire the ability to invade the lining of the intestine wall, causing ulcers and a condition known as balantidial dysentery, which is similar in its symptoms and its effects to amoebic dysentery. Severe cases can prove fatal.

Two particularly interesting species are *Opalina ranarum* and *Opalina obtrigonoidea,* which are parasites of various species of frogs and toads respectively. Throughout most of the year frogs and toads live a mainly terrestrial existence, during which their reproductive organs are inactive and they are incapable of breeding. At this time, too, the parasites exist in the hosts' tissues in a non-reproducing phase, occasionally perhaps dividing by simple fission. With the approach of spring, however, the amphibians' reproductive organs become active and increase in size as production of eggs and spermatozoa begins, in preparation for the breeding season when the frogs and toads migrate to the water to mate. This activity is accompanied by the production of sex hormones, which are released into the blood. These hormones exert an effect on the parasites, which now begin to divide rapidly, producing many smaller forms that proceed to cover themselves each with a cyst wall. Each cyst usually contains four, though sometimes more, nuclei. After the amphibia have copulated, the *Opalina* cysts pass out into the water via their faeces.

They remain dormant in the water, and indeed are incapable of further development unless they are swallowed by the tadpoles which hatch after a few days from the amphibian eggs. Once in the tadpole's stomach, however, the cyst wall is dissolved away to release the small multinucleate cell contained within it. By further nuclear division, followed by division of the cytoplasm, each former cyst gives rise to a number of male or female gametes. These now fuse together in pairs to form zygotes, which in turn develop into adults by about the time the tadpoles are ready to metamorphose into small frogs. Thus it is that the young frogs obtain the parasites which will remain virtually dormant in their bodies until they eventually approach their first breeding season, when the sequence of events just described begins all over again.

One other ciliate must be mentioned—*Ichthyophthirius multifilius*—which is a well known parasite of acquarium fish, in which it causes a disease known as 'white spot' that often kills the fish unless it can be satisfactorily treated. *Ichthyophthirius* must be regarded as an internal parasite, because it penetrates beneath the hosts' skin, where it forms the characteristic white spot, within which the original single individual divides into two or four individuals. When these have grown to full size, they work their way out through the skin and sink to the bottom of the aquarium tank, where they encyst. The contents of each now divide until the cyst wall contains between 100 and 1,000 minute individuals. These now break out of the cyst wall as actively swimming young ciliates, which move around in search of a fish host, beneath whose skin they can grow to full size and thus repeat the cycle.

CHAPTER SEVEN

INTRODUCING WORMS
AS PARASITES: THE FLUKES

Although the term 'worm' is commonly used, and most of us think we know what the word means, in fact it has no precise biological definition. There are in fact a number of quite distinct and unrelated phyla whose members are commonly referred to as worms, and no fewer than four of these have members which lead a parasitic mode of life. It is the parasitic members of these four phyla that we shall be considering in this and the next two chapters.

Biologically the most primitive of the four phyla is the phylum Platyhelminthes, the members of which are commonly referred to as flatworms, for the very good reason that their bodies are extremely flattened from above. There are three distinct classes: the Turbellaria, the majority of whose members are free-living; the Trematoda, the flukes, all of which are parasites; and the Cestoda, the tapeworms, which are also all parasites. The phylum Nematoda comprises the roundworms, many of which are important parasites of many kinds of animals and plants, but which also includes many free-living kinds as well. The phylum Acanthocephala is a small group of relatively unknown parasitic animals usually known as the spiny-headed worms, whose adults live as parasites in the intestines of vertebrates. The members of the phylum Annelida are the true or segmented worms, easily recognisable because the adult body is prominently ringed, the rings or annuli indicating the limits of the internal segments into which the body is divided. There are three classes—the Oligochaeta, which are the earthworms;

63

the Polychaeta or marine worms; and the Hirudinea, which are the leeches, and the only parasitic members of the phylum.

Although broadly similar in general appearance, the parasitic trematodes differ in certain important respects from the free-living turbellarians. The body of the latter is covered with cilia whose rhythmical beating enables the worms to glide along on the bottom of the pond or stream in which they are living. The trematodes have lost these cilia and, therefore, the ability to move, at least in the adult, though, as we shall see, some of the larval stages are ciliated. The outer body layer or cuticle is much better developed in the trematodes than it is in the turbellarians, and the trematodes have suckers at the front and hind ends of the body which enable them to adhere to their hosts. Nearly all flatworms, whether free-living or parasitic, are hermaphrodite, with extremely complicated reproductive systems, but the arrangements for mating often preclude self-fertilisation.

The class Trematoda is divided into three orders. The members of the order Monogenea are mainly external parasites, many of them living on the gills of fishes. They have simple life cycles without a number of larval stages. At the hind end they usually have a number of adhesive suckers and hooks with which to maintain their hold on the host, and often another sucker surrounding the mouth. In the order Aspidogastrea the whole of the ventral surface forms an adhesive organ consisting of rows of tiny suckers. The members of this group are mainly internal parasites of certain vertebrates, especially fish and turtles, and of molluscs and crustaceans. They are not found in warm-blooded animals. Their life cycles are simple. The vast majority of the Trematoda, however, belong to the third order, the Digenea. These are the true flukes, and include many important parasites of domestic animals and man. There is a single ventral sucker which is the main adhesive organ, and a well developed sucker surrounding the mouth. They have complicated life histories, usually involving an invertebrate host for one or more of the larval stages and a main vertebrate host for the adult stage.

The best known member of the Monogenea is *Polystoma integerrimum,* which is not a gill parasite but lives in the adult stage in the bladders of various kinds of European frogs. It has a highly developed system of suckers to enable it to maintain its hold on the bladder wall. The under surface of the hind end of the body consists of a muscular disc on which there is a semi-circle of six suckers, with a pair of hooks just in front of the middle two. At the extreme front of the body there is another sucker surrounding the terminal mouth.

The life history of Polystoma is of particular interest because it is the only known flatworm whose reproductive cycle is linked with the reproductive cycle of its host. During the winter, when it is in hibernation, the frog's reproductive organs develop and become mature by the time it is ready to come out of hibernation and migrate to its home pond to breed. At this time, too, the reproductive organs of *Polystoma* also become mature. When the frogs reach their breeding pond, they come together in pairs; after an interval the female sheds her eggs into the water while the male pours his seminal fluid over them. This is the signal for the *Polystoma* in the frogs' bladders to shed their vast quantities of eggs, which are passed out into the water when next the frogs urinate.

Within a few days the frogs' eggs hatch to tiny tadpoles with external feathery gills. After a few days these gills wither and are replaced by internal gills. The rate at which the *Polystoma* eggs develop is so regulated that when they hatch, the tadpoles have already reached the internal gill stage. These gyrodactylid larvae, as they are called, are barrel-shaped, and have one large posterior sucker armed with sixteen hooks and five incomplete circles of cilia around the body. As soon as they are hatched, they seek out a tadpole and enter the gill chamber, where they attach themselves to the internal gills and begin feeding on the mucus which covers the gill filaments.

Eventually the tadpole begins the metamorphosis which will convert it to a young frog, and one of the changes which takes place at this time is the absorption of the gills as they are replaced by lungs. When this happens, the gyrodactylid larvae leave the disappearing gills, pass through the young frog's gut and make for the bladder. During this journey they lose their cilia and larval hooks, and develop their adult suckers and hooks.

Sometimes something goes wrong with the timing of the *Polystoma* egg development, and the gyrodactylid larvae hatch too soon, while the tadpoles still have their external gills, so it is to these and not the internal gills that they become attached. This has a profound effect on their development. Within 2–3 weeks they become premature adults, but many of their organs are rudimentary, even their reproductive organs. These can, however, yield reproductive products which give rise to quite normal gyrodactylid larvae. By the time these larvae are ready, there will still be some tadpoles with internal gills to which they become attached and undergo normal further development.

Experimental work has shown that the onset of sexual maturity in *Polystoma* is controlled by hormones released in the normal course of events by the frog's pituitary. Experimental injections of pituitary extract into a frog known to be infected with *Polystoma* cause the flatworm to become sexually mature within 4–8 days and then to produce quantities of fertilised eggs over the next week or so. Normally, however, the development of *Polystoma* once it has reached the frog's bladder is very slow, and it requires 3 years to reach sexual maturity. By this time the frog will also be full grown and ready to breed.

As mentioned already, the majority of the Monogenea live as external parasites on the gills of fishes, feeding on blood which they extract from the gill filaments. An interesting examination was carried out at Plymouth from 1953 to 1955 on the occurrence of these parasites on various fish common in the area; the results were very interesting, showing among other things that in many species more than half the population were subject to the attentions of the parasites. Perhaps the most astonishing result was that virtually every mackeral examined had at least one member of its own particular parasite species, *Kuhnia scombri*. The survey also revealed that some of the parasitic species are very particular about the precise position on the host they preferred : for instance, *Diclidophora merlangi,* a parasite of the whiting *(Gadus merlangus),* nearly always settled on the front gill of the fish, whereas the closely related *Diclidophora luscae,* a parasite of a relative of the whiting called the pout *(Gadus luscus),* appeared almost exclusively on the second and third gills, hardly ever on the front gills.

Two other species must be mentioned as being of particular interest. *Gyrodactylus elegans* is a small species which lives as an ectoparasite on a number of small marine and freshwater fish. It is viviparous, the single larva developing within the uterus and coming eventually to occupy most of the bulk of the parasite. What is particularly interesting is that in the uterus of the developing larva another larva is already developing, representing the next generation, and even this may have the rudiment of yet another larval generation developing within its embryo uterus, the whole thus forming a kind of Chinese box. As many as four generations have been identified in one single individual.

Diplozoön paradoxum is a quite remarkable animal, a common parasite on the gills of the minnow. What appear to be one animal when taken from the gills is in fact two individuals which have become one. The attachment between the two takes place when they are larvae. At first they adhere to each other by an arrangement of suckers, but these disappear when the two bodies actually become fused together. Although joined together, each individual has a complete set of internal organs, including reproductive organs, and it is believed that cross-fertilisation between them is the rule. So fixed is this dual pattern that any individual larva which fails to find a partner cannot go on living, and dies without attaining sexual maturity.

The second trematode order, the Aspidogastrea, contains only a relatively small number of species. As already mentioned, the members of this group are distinguished by having the whole of the under surface composed of rows of rectangular suckers. The species whose life cycle is best known is *Aspidogaster conchicola,* which lives in the bodies of freshwater mussels belonging to the genera *Anodonta* and *Unio.* The larvae develop inside the eggs while they are still within the uterus of the parent, and they may hatch out while still in the host or pass out of the parent body into the surrounding water. If an infected mussel containing the parasite and its eggs is eaten by a fish, it is possible for the larvae resulting from these eggs to attach themselves to the wall of the stomach or the intestine of the fish, which thus becomes a second host for the parasite.

Perhaps we have here a hint of how the more complicated life histories of the Digenea, in which there are always at least two hosts in the complete life cycle, might have arisen. From the point of view of parasitism, in fact, the most important of the trematodes are the members of the order Digenea. The adults live in the gut of vertebrates or its associated glands, such as the liver, bile duct, gall bladder, lungs, pancreas or bladder. The sucker arrangement in these adults is simpler than it is in the other two orders. There is a single ventral sucker on the under surface of the body, and an oral sucker at the front end surrounding the mouth.

The majority of species are hermaphrodite, though in one important family, the Schistosomatidae, the sexes are separate. But without exception the life cycle is complicated, involving at least one intermediate stage in which the larvae live; and there are always a number of different larval stages between one adult stage and the next.

The best known and one of the most important of all these trematode parasites is the liver fluke, *Fasciola hepatica,* and an account of its life history will show clearly the roles of the various trematode larval stages. The adult liver flukes live in the liver and bile ducts of sheep, goats and cattle, causing the disease known as liver rot, which can be fatal. The body of the fluke is leaf-shaped and extremely flattened, about 1in long and $\frac{1}{2}$in wide. Each adult produces a large number of eggs which, when they are released, pass down the bile duct into the intestine to reach the exterior with the host's faeces.

Trematode eggs cannot withstand desiccation, as those of some other parasites can, and only those which are deposited in or beside a pond or some other area of water can survive, each then hatching out to form an active ciliated miracidium larva, whose purpose is to swim around and search for a member of the intermediate host species. In Britain this is the dwarf pond snail, *Limnea truncatula,* but *Fasciola* is extremely widespread, and in different parts of the world it uses different intermediate hosts; no fewer than thirteen different species of *Limnea* serve this purpose, while in the USA one of four species of *Galba* are used. Since the vast majority of the eggs will not be deposited near enough to water to hatch, they will perish; and of the minority which do manage to hatch only a small proportion of the resulting miracidia will be able to find a pond snail in the short time they are able to live. But the relatively small number of larvae which do finally find sanctuary in a snail make up for these losses by eventually producing an enormous number of offspring by the end of their very complicated life history.

The successful miracidium makes for the snail's lung, where it loses its cilia and turns into a second type of larva known as a sporocyst. This is little more than a sac within which virtually all its body material is reorganised to form a number of separate larvae, called rediae, which are released by the bursting of the sporocyst wall when they are fully formed. As soon as they are released, these rediae penetrate the wall of the snail's lung to bury themselves deeper in the body, usually ending up in its liver.

What happens to the rediae now depends upon the time of year. In the late summer or autumn each redia will divide internally to produce a second generation of rediae, and these may even give rise to a third generation during the winter. But in the spring and summer each redia divides internally to produce the fourth type of larva, the cercaria, a rather tadpole-like creature with a long tail, capable of swimming. These cercariae leave the snail's liver and work their way out of its body into the pond. They swim to the edge of the pond, come out of the water and climb to the top of a grass blade. Here they shed their tails and cover their bodies with a hard resistant coat which protects them from desiccation. Although, as already mentioned,

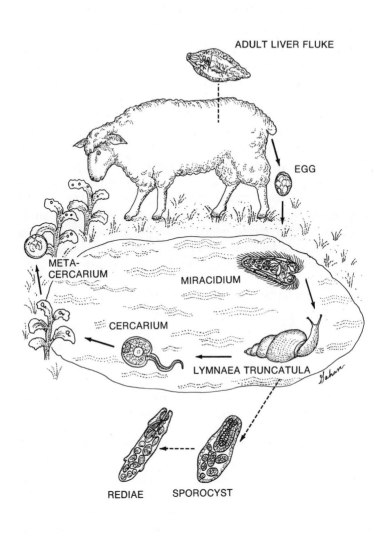

ADULT LIVER FLUKE

EGG

META-
CERCARIUM

MIRACIDIUM

CERCARIUM

LYMNAEA TRUNCATULA

REDIAE   SPOROCYST

Fasciola hepatica (liver fluke) cycle

only a small proportion of the large numbers of eggs produced by each adult liver fluke eventually give rise to miracidia, each miracidium which does succeed in locating a snail host can eventually be responsible for the production of something like 600 cercariae.

Although the encysted cercariae on the tops of their grass blades can survive for a long time, their further development is dependent upon their being eaten by a sheep or a cow. When they reach its duodenum, the hard cyst wall is dissolved away by the digestive enzymes and they are able to work their way through the wall of the duodenum into the body cavity, from which they eventually manage to get to the liver. Here they soon become transformed into young adult flukes. Man very seldom becomes infected with *Fasciola hepatica,* although it can live and thrive in his liver, since he does not eat the grass blades on which the cercariae are normally encysted. Occasionally, however, humans are infected by eating watercress on which cercariae have encysted.

One relative of *Fasciola* which is parasitic in man is the Chinese liver fluke, *Clonorchis sinensis.* Besides its human importance it is interesting because it has not one but two intermediate hosts, in addition to its main host. It is one of a group of related parasites whose adult stages live in fish-eating mammals. *Clonorchis* is common and widespread among man, cats and dogs in China, Japan and neighbouring countries of the Far East.

The adult flukes live in the bile ducts of their mammalian hosts and their eggs pass out with their hosts' faeces. Their chances of reaching water are fairly good, bearing in mind the general sanitary arrangements in these eastern countries. Within these eggs miracidia develop, but they do not emerge from the egg case as they do in *Fasciola.* Instead the eggs are eaten by one of three species of water snail. Within the snail the miracidia hatch and develop into sporocysts, which in their turn produce a number of rediae. These also multiply to produce cercariae, which are now released into the water.

The behaviour of *Clonorchis* cercariae is quite different from those of *Fasciola*. They seek out freshwater fish, burrow into their skin, shed their tails and encyst by forming a resistant coat around themselves. Here they remain until the fish is caught, perhaps by a cat, perhaps by man; in either case the fish is likely to be eaten raw, and then the cercariae break out of their cysts in the stomach of whoever has eaten the raw fish. They now make their way to the liver of their final host to begin the complicated life cycle all over again.

The control of human *Clonorchis* infection is really quite simple. If all freshwater fish were cooked before being eaten, the fluke could be rapidly wiped out so far as human hosts were concerned. But so ingrained is the habit of eating raw fish that it would need an enormous educational campaign to have any hope of breaking it. It is said that in parts of Southern China as many as 75 per cent of the inhabitants are infected. To eliminate the fluke altogether would be a much more difficult task, because it also infects cats, which of course eat raw fish and thus constitute a reservoir host.

A number of different species live as adults in various water birds and have two intermediate hosts. The first of these is usually a water snail, and the second a fish or another mollusc of species commonly eaten by gulls and other water birds. Adults of the species *Cryptocotyle lingua* spend their lives attached to the intestinal walls of a great variety of coastal birds, including most gulls, terns, razorbills and kittiwakes. The vast quantities of eggs they produce pass into the sea with the birds' droppings. In about 10 days they have each produced a miracidium larva, but this remains within the egg until that is eaten by a common periwinkle, *Littorina littorea*. From the gut of the winkle the miracidium migrates into its body, usually coming to rest in its liver, where it causes a certain amount of destruction. An interesting method of deciding whether an individual *Littorina* is infected or not is to look at its foot, for liver pigment is released if part of the liver is destroyed by the parasite, and this pigment colours the muscle forming the snail's foot. If its foot is coloured, the winkle is infected.

In the liver the miracidium becomes converted to a sporocyst, which in its turn gives rise to rediae, and these again to cercariae. These cercariae now make their way through the tissues of the snail and liberate themselves into the sea, where they swim around looking for almost any kind of shore fish. They can penetrate almost any part of the fish's body, but seem to prefer the fin rays. Wherever they penetrate, they remain just below the surface and proceed to secrete a cyst wall around themselves. The tissues of the fish, in response to the irritation cause by the presence of the cyst, cover the original cyst with a second wall of tissue. Of course only a minute proportion of the cercariae will succeed in finding the second host, the majority of them dying without succeeding. But enormous numbers of them are produced. It has been estimated that a single periwinkle can release as many as 3,000 cercariae in a day.

Any subsequent development depends upon the fish being eaten by a seabird. When this happens, only the outer cyst coat, which was secreted by the fish, is dissolved by the stomach juices of the bird; the inner coat, secreted by the cercaria itself, is only dissolved away when it reaches the intestine. Once released, the cercaria loses its tail, changes to a young adult fluke and attaches itself to the intestinal wall.

Parorchis acanthus, which is also a parasite of gulls and terns, has a somewhat different life cycle. The eggs develop while they are still in the adult fluke's uterus, so that when they pass out of the body of the host bird, the single miracidium in each egg is fully developed and is released into the water almost immediately. There is no sporocyst stage. Within the body of the miracidium a single redia larva develops. Further development depends upon the miracidium coming in contact with a carnivorous member of the snail family, the dog whelk, *Nucella lapillus.* Once inside the snail's body, the single redia is released and proceeds to divide and develop to form a number of cercariae, which leave the body of the snail.

As their second hosts, these cercariae seek out bivalve molluscs, either the common mussel, *Mytilus edulis,* or various cockle species belonging to the genus *Cardium.* Here the cerariae encyst, and further development into adult flukes can only take place if the cockle or the mussel is eaten by a seabird.

We turn now to members of the family Schistosomatidae, the blood flukes, whose members show two important differences from all the other trematodes. The adults of all species exist as males and females, and live in the blood vessels of mammals or birds. They are much more worm-like than other trematodes, the body of the male being rolled ventrally to form a kind of canal in which the thinner body of the female is enveloped. Oral and ventral suckers are present, the latter being better developed in the male than in the female, since it is his responsibility to maintain a hold on the wall of the blood vessels so that the pair are not swept along in the continuous current of blood. The eggs of all members of the group are provided with a spine, which enables them to work their way out of the blood vessels in which they are laid and eventually reach the cavity of the gut, and then the exterior via the host's faeces.

Three species of blood flukes are parasitic in man, the most important and widespread being *Schistosomum japonicum,* which is common in China and Japan and throughout the Far East. The adult blood flukes live in the veins of the intestinal wall, from which the eggs work their way into the cavity of

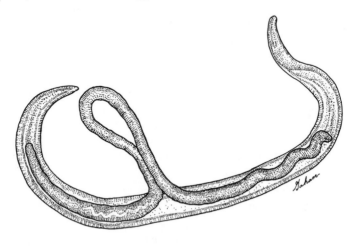

Schistosoma mansoni (blood fluke) copulating

the intestine and thence the exterior. With a modern sewage system, this would be the end of them, but in these far eastern countries human faeces are valuable for fertilising the ground. Hence the eggs gain ready access to the paddy fields, and are able to hatch as active miracidia. These swim around until they have found one of the water snails that are very common in these surroundings, which they use as intermediate hosts.

In the body of the water snail the miracidia lose their cilia and develop into sporocysts. The redia stage is missed out, each sporocyst giving rise directly to a number of cercariae, which make their way out of the snail's body and swim about near the surface of the water. As soon as a cercaria comes in contact with a leg of someone tending the rice fields, it is able to attach itself to the skin and work its way into a blood vessel. Once installed, it allows itself to be carried around until it reaches the veins in the wall of the host's intestine, where it is rapidly transformed into either a female or a male fluke. These then come together in pairs and the cycle begins all over again. As with many other flukes, the rate of egg production is considerable, one pair of blood flukes being credited with as many as 500 per day indefinitely.

The other two human blood flukes are *Schistosoma mansoni,* which is a widespread species, occurring in most parts of Africa, Madagascar, the West Indies and South America; and *Schistosoma haematobium,* which is mainly an African species, but also occurs in Mauritius and Madagascar. Their life histories are essentially similar to that of *japonicum,* the intermediate hosts also being various species of water snail.

CHAPTER EIGHT

TAPEWORMS

Apart from the fact that their bodies are also flattened, there is little about the structure of the tapeworms to suggest that they are related to the turbellarians and the trematodes. They are more completely adapted to a parasitic mode of life than the trematodes. Having no mouth or digestive system, the adults can only survive in tubular parts of the host's body where they are surrounded by liquid from which their food requirements can be obtained by diffusion through their outer surface. The small intestine is of course the ideal site, since here the tapeworm will be constantly surrounded by the soluble products of the host's digestion; and the vast majority of tapeworm species anchor themselves to the wall of the intestine as adults. A few species are found in the bile duct and the pancreatic duct.

The two best known human tapeworms are the beef tapeworm, *Taenia saginata,* which commonly occurs in all beefeating countries, and the pork tapeworm, *Taenia solium.* Until fairly recently the pork tapeworm was much more common than the beef tapeworm, which explains why it is featured in most biology textbooks, but it has been disappearing fast, and in some countries where it was quite common is thought now to have become almost extinct; at the same time the beef tapeworm has become more common.

An account of the structure and life history of the beef tapeworm will serve as a guide to tapeworms generally. The front end of the adult tapeworm consists of a minute head or scolex by which it anchors itself to the wall of the host's intestine so that it cannot be swept away by the current of digested food

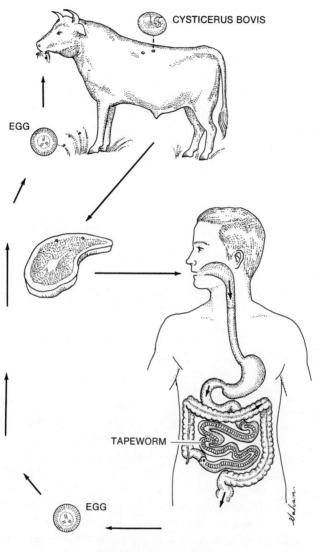

Beef tapeworm cycle

which is continually passing through. For this purpose it is provided with four muscular suckers. The scolex of *Taenia solium* and of some other species also has a circle of two rows of curved hooks forming a rostellum, which reinforces the actions of the four suckers.

Behind the scolex the remainder of the extremely long body consists of a large number of separate segments known as proglottids, and more of these are continually being budded off from the hind end of the scolex as it grows. Each proglottid is virtually a separate individual, producing a complete set of male and female reproductive organs, for its function is to produce vast numbers of eggs.

After the eggs within a particular proglottid have been fertilised, all the other structures in the proglottid degenerate, so that it becomes little more than a complicated branched uterus packed with ripening or ripe eggs. As new young proglottids are budded off from the hind end of the scolex, proglottids whose eggs are completely ripe are shed at the hind end of the tapeworm itself, and pass down the remainder of the host's gut to be released from the body with the faeces. In this way, once a tapeworm has become full grown, it does not necessarily continue to increase in length, proglottids at the hind end being shed at the same kind of rate at which new ones are budded off from the scolex.

The average length of a full grown tapeworm depends very much upon the species to which it belongs. The average length of *Taenia saginata* is 12–24ft, but specimens up to 50ft in length are not uncommon. The record authenticated length was about 82ft. Because of their length, *saginata* specimens are usually doubled back on themselves in the host intestine, and there is cross-fertilisation between the proglottids, the penis of one proglottid being inserted in the vagina of another one. It is not uncommon for one host to harbour more than one tapeworm, and in this case cross-fertilisation between two individuals may well occur. A full grown specimen may consist of up to 1,000 proglottids, each producing as many as 100,000 eggs, and up to ten proglottids may become detached and reach the exterior with the host's faeces each day.

Development of the eggs cannot take place unless they are eaten by an intermediate host. In the USA and in Europe cattle are the normal intermediate hosts, but in tropical countries other ruminants such as goats, sheep, llamas and even giraffes serve equally well. In the intermediate host's stomach the protective shell covering the egg is dissolved away by the digestive enzymes to release the first stage embryo, which by this time has formed within the egg. This is known as a hexacanth embryo or onchosphere, and has a circle of six curved hooks at one end with which it is able to penetrate the wall of the stomach and so reach a blood vessel. It allows itself to be carried round in the blood until it arrives in the blood capillaries of a muscle, where it brings its hooks into play once more to prevent further progress through the blood vessel.

It now bores its way through the capillary wall and settles down between the muscle fibres, losing its hooks and changing into a fluid-filled bladder—the bladder-worm or cysticercus. From one point on its wall a scolex develops. This projects inwards so that it is inside-out like the finger of a glove pushed inwards towards the palm.

Cysticerci can remain alive in this position for many months, but further development can only take place if they reach the intestine of their main host, man. This can happen if a joint of beef or a steak containing the embryos is eaten without being sufficiently well cooked to kill them; and of course underdone beef and rare steaks are extremely popular. When one of these bladder-worms which has escaped being killed by cooking and being crushed by man's teeth reaches his small intestine, the scolex becomes everted, so that it now projects from the surface of the bladder the right way out. The bladder is shed and the young scolex immediately attaches itself to the intestine wall. Here it soon begins to grow and bud off the first proglottids as it absorbs the food materials which are continually passing by it.

One problem to consider is how the cattle or other intermediate hosts become infected with the beef tapeworm eggs. In the country, of course, it may sometimes happen that human faeces laden with tapeworm eggs may be deposited behind a hedge, where the proglottids may then be picked up by a cow, but the rate of incidence of the tapeworm in civilised countries is much too high to be explained in this way. The real reasons are undoubtedly connected with the methods of sewage disposal and the use of treated sewage sludge as fertiliser.

Tapeworm eggs are extremely resistant to all kinds of treatment, and can survive all the normal processes used to treat sewage in urban sewage works. If such treated sewage, in which harmful bacteria will have been killed, is then used on fields, the tapeworm eggs are still active. They are still capable of developing if they are picked up by grazing cattle, which is of course quite likely. In many places raw sewage is discharged into the sea, and from this many tapeworm eggs will be washed up on the shore, to be eaten by gulls and other birds which feed there. The tapeworm eggs will pass right through the birds' bodies unharmed, and since many of these seabirds have a habit of moving inland to roost at night, they will spread the still living tapeworm eggs on the land in their droppings.

Taenia solium is a smaller tapeworm than *saginata,* averaging 9–15ft in length. Its life history is very similar to that of *saginata,* except that the pig is the intermediate host.

In addition to the two species of *Taenia,* there is one other important human tapeworm, *Diphyllobothrium latum,* usually known as the broad tapeworm because the proglottids are wider than they are long, in contrast to the mature proglottids of *Taenia,* which are longer than they are wide. It belongs to a group of tapeworms which have two intermediate hosts, and its larval stages are different from those of the majority of tapeworms.

Diphyllobothrium latum is sometimes known as the fish tapeworm, because its second intermediate host is always a fish, and therefore the adult tapeworm is found only in fish-eating mammals. Besides man these include cats, dogs, bears, seals and porpoises. And in man it is particularly common in countries where fish is eaten either raw or partly cooked, and this includes smoked. It is in fact a very widely distributed parasite, occurring in Australia, Japan, Siberia, many parts of central Europe, the Near East, Central Africa, Canada, the USA and parts of South America. It is similar in size to the beef tapeworm, and therefore larger than the pork tapeworm. Multiple infection can occur, and there is a gruesome reference to a Russian woman who harboured no fewer than six specimens with a total length of 290ft. She must have eaten an enormous amount of food to have kept these going as well as herself.

Ripe proglottids are regularly shed from the hind end of *Diphyllobothrium,* as they are from *Taenia,* and pass out with the host's faeces. Since the host is a fish-eating mammal, the chances of the faeces being dropped in water are quite good, and those proglottids which find themselves in water soon release their eggs. Meanwhile development has been proceeding in the egg, so that by the time it is released, it already has inside it a fully formed ciliated larva known as a coracidium. This is able to swim around, but is unable to seek out its required host, which is a small aquatic crustacean, usually a copepod. Copepods are on the lookout for coracidia, however, and before long the larva is likely to be swallowed by its first host.

Inside the coracidium meanwhile a normal hexacanth embryo or onchosphere with six hooks has developed, and as soon as the coracidium finds itself in the gut of the crustacean, it sheds its ciliated coat to release the onchosphere. This works its way through the gut wall until it reaches the crustacean's body cavity, and here in about 3 weeks it develops into the next stage larva, known as a procercoid, which still has the six hooks at one end of the body.

No further development can take place unless the copepod is swallowed by a fish, again quite a likely event, since there are many plankton-eating fish. In the gut of the fish the procercoid sheds its ring of hooks, bores its way through the gut wall and wanders about the body. Eventually it will settle, usually in the liver or the muscles, and for the next 6 weeks gradually changes to the last larval stage, the plerocercoid. This now waits for its present host to be eaten by its final mammalian host, and when this happens, it is ready to begin its adult life, for it possesses a well formed scolex.

If its intermediate fish host is eaten by another and larger fish, the plerocercoid finds itself in the gut of a second fish host, whence it makes its way through the wall of the gut to secrete itself in the liver or muscles. The second fish host is what is known as a paratenic host, that is an additional host in which no further larval development occurs.

CHAPTER NINE

ROUNDWORMS AND LEECHES

The nematodes or roundworms constituting the phylum Nematoda may not be so well known as some other groups of animals containing important parasites, but some of the members of the phylum are very serious parasites both of man and other animals. It is a large group, many of which are free-living both in the soil and in water. The majority of them are small, and many of them microscopic. Only about a dozen or so are serious parasites of man, but large numbers of minute forms are serious parasites of plants, living among and destroying their tissues.

Whatever their size, nematodes are remarkably uniform in appearance, all of them being slender and elongated. The name nematode derives from the Greek word *nema,* meaning a thread. The mouth and the anus are at the extreme ends of the body, the surface of which is covered with a resistant transparent cuticle. In most species the sexes are separate, and the eggs are covered with a tough resistant coat.

Like the arthropods, the larvae of the nematodes develop through a series of moults or ecdyses, and so far as is known there are always four of these moults before the adult stage emerges. In many species the first two moults take place while the developing larva is still enclosed within the original eggshell, so that the larva which finally emerges has to burst through two cast larval skins as well as the shell. There is considerable variety in the life histories of the parasitic members of the phylum. In many cases there is an intermediate as well as a final host.

Five different kinds of nematodes are serious parasites of man. Our first example is *Trichinella spiralis,* which causes the disease known as trichinosis, associated with eating insufficiently cooked pork. Whereas with most parasite diseases it is the adult parasites which cause the damage, in *Trichinella* the adults are relatively harmless, the disease being caused by the larvae.

Infected pork contains last stage larvae, which are released in the duodenum if they have not been killed during the cooking. In 2–3 days they will have moulted for the last time and given rise to adults which become sexually mature almost immediately and are ready to mate. They are tiny worms, the females being about 4mm in length and the males about $1\frac{1}{2}$mm. After mating, the males soon die, but the females, which are viviparous, embed themselves in the wall of the duodenum. Here the 1,500 or so fertilised eggs in the uterus proceed to develop and moult to release first stage larvae, which are then born. These tiny larvae, not more than $\frac{1}{10}$mm long, work their way into the blood vessels and are then carried round the body.

They leave the blood vessels to enter the muscles in various parts of the body, but seem to prefer the muscles of the diaphragm, tongue and ribs, though many of them will also be found in the larger muscles of the arms and legs. Once installed, they proceed to grow until they have reached a length of about 1mm, or ten times their original size. Growth now ceases, and each larva becomes enclosed in a thick-walled cyst produced by its host's tissue cells. Further development is impossible unless the flesh is eaten, which will not of course normally occur in a human infection. But besides man there are a number of other animals which are susceptible to *Trichinella* infection, as we shall see.

Adult worms in the intestine do little harm. The danger period is when the enormous numbers of larvae are migrating round the body in the blood and invading the muscles. A heavy infection can at this time cause extremely acute muscular pains, fever and anaemia. If death is going to occur, it will take place during this migratory phase. Once the larvae have encysted in the muscles, the symptoms subside and the danger is past, though the cysts may cause some permanent damage to the muscles.

The majority of human parasites tend to be much more common in tropical and subtropical regions than they are in temperate areas, but *Trichinella* is essentially a parasite of temperate and cold climates, being found mostly in North America, Europe and in the Arctic, where it is a common parasite of the Eskimos. Here it is also found in polar bears, dogs, seals, walruses and white whales, and it is from the seals, walruses and whales, which form an important part of the Eskimos' diet, that they obtain their trichinosis infections. It has been suggested that the Jewish laws forbidding the eating of pork may have been an attempt to prevent trichinosis, but this seems unlikely, because the Near East is too far south for *Trichinella* to have flourished.

Only a small proportion of those infected with *Trichinella* are diagnosed as suffering from trichinosis. Presumably the majority of infections are light and give only mild symptoms, which lead them to be diagnosed as some minor illness. In one particular survey, post-mortem examinations of more than 200 bodies showed them to contain encysted *Trichinella* larvae, yet not a single one of these had been diagnosed as suffering from the disease during life. It is known that infection induces some degree of immunity, which might explain the minimal impact of these infections. It has been estimated that as many as 16 per cent of the population of the USA are infected with *Trichinella,* though in the vast majority of cases the infection is never suspected.

The vast majority of human infections result from eating infected pork, either directly as meat or in the form of sausages made from the meat. How are the pigs infected? Certainly they get no opportunity of eating infected human flesh, and no eggs are passed out by infected humans or any other animals. The answer lies in the methods generally adopted for feeding pigs on swill or garbage. The kitchen waste collected for pig feeding may well contain scraps of pork which has not been cooked sufficiently to kill the encysted worms, and these may still not be killed by the methods used to prepare the waste before it is fed to the pigs. In some countries, such as Great Britain and

Canada, pig breeders are required to boil all swill for at least ½hr before feeding it to their pigs, and this is sufficient to kill any larvae which might be present.

Rats may also be responsible for pig infections, for they are very prone themselves to *Trichinella* infection. If an infected rat appears among the pigs, it is very likely to be killed and eaten, thus infecting the pig.

More people become infected with trichinosis through eating sausages containing infected pork meat than through eating joints of pork. It has been estimated that 1oz of heavily infected sausage meat may contain as many as 100,000 encysted larvae. The dangers of trichinosis infection are greatest among Germans, Italians and Austrians, through their fondness for all kinds of smoked and other uncooked sausages. Many of the pigs used to provide the meat for these sausages are home-killed, and no precautions are taken to kill any encysted *Trichinella* larvae the meat may contain. It is said that in these countries schoolteachers and country priests are particularly exposed to the risks of trichinosis infection, for they are often invited in to taste the latest batch of home-cooked sausage!

Whereas animals infected with most parasites are usually fairly easy to detect, those infected with trichinosis are not. Even the most stringent examination of carcases will not detect every one harbouring encysted larvae. There are, however, two measures, either of which, if adhered to, could virtually eliminate the disease. If all pork was sufficiently cooked, all the larvae would be killed. Alternatively, storage at $-15°C$ for 48hr, while it will not kill the encysted larvae, will render them incapable of developing and reproducing should they subsequently reach a potential victim's stomach without being killed during the cooking.

Two relatives of *Trichinella* that deserve mention are the giant kidney worm, *Dioctophyma renale,* and the so-called whipworms belonging to the genus *Trichuris.* These are mainly parasites of various mammals, but they are also sometimes found in man.

Dioctophyma is one of the giants among the nematodes, the

females averaging about 2ft in length, with exceptional individuals exceeding 3ft, and the smaller males about 1ft in length. The most common hosts are otters, mink and other fish-eating carnivores, as well as dogs, foxes and occasionally man.

Infection begins with the ingestion of last stage larvae, which soon moult to produce young adults. These make for the host's kidneys, where they bury themselves and begin to feed on the kidney tissues. By the time the worm has grown to full size it will have reduced the kidney to a functionless shell. For some unexplained reason only the right kidney is usually invaded, the left one growing to about twice its normal size to enable it to carry out the functions of both kidneys. Occasionally the worms will settle in the general body cavity instead of the kidneys. In some countries, notably Canada, where large-scale mink farming is undertaken, *Dioctophyma* infection can assume economic importance owing to the devastation it causes among the breeding stocks.

The life history of the kidney worm is quite complex, involving no fewer than three intermediate hosts. Ripe eggs can only develop if they are shed into freshwater and eaten by an annelid worm, *Cambarinocola chirocephala,* which lives in partnership with freshwater crayfish. The first stage larvae, which hatch in the gut of the worm, make their way out of its body and invade the tissues of the crayfish, where they become encysted.

Further development can only take place if the crayfish is eaten by a freshwater fish, in whose gut the cyst walls are dissolved to release the larvae. Growth and moulting now continue until the last larval stage is reached. The final moult, which will produce the adults, can only occur if the fish is eaten by a mink or one of the other final hosts.

Whipworms are so called because the long anterior part of the body is thin and whip-like, in contrast to the shorter and much thicker hind end containing the reproductive organs. When they were originally discovered it was thought that the thin anterior end was in fact the tail, and so they were given the generic name *Trichuris,* which means 'thread tail'. When the truth about them was finally discovered, it was suggested that the name should be changed to *Trichocephalus,* which means 'thread head', but this was impossible because the rules of nomenclature in biology decree that the name first given to an animal carries priority over all subsequently proposed names. So they are still thread tails.

Adult whipworms are gut parasites found in the lower intestine of a wide variety of mammals, including dogs, various rodents and ruminants, monkeys and man. Infection is not usually very serious unless the number of worms reaches enormous proportions. They are quite small worms, averaging 1–2in in length. The human species, *Trichuris trichiura,* is a widespread and common parasite in many parts of the world where the climate is warm and moist.

The life cycle is simple, the fertilised eggs passing out with the faeces. They are very susceptible to desiccation, and will soon die unless they are deposited on permanently damp or wet ground. In any case development is slow, and even under moist warm conditions it takes 3–6 weeks for the embryos to develop to the point of hatching. The incidence of *Trichuris* infection depends very much upon the prevailing conditions of hygiene and sanitation. In many parts of the world native children will defaecate on the ground surrounding native huts, and in consequence huge numbers of infective eggs containing fully developed embryos will swarm in the soil. If fingers touch this soil and then pick up food without being washed, or are brought in contact with the mouth, infection will result. Whipworm infection is comparatively rare in countries with high standards of hygiene and sanitation.

The second important group of nematodes are the ascarids and their relatives the pinworms and *Strongyloides.* The best known member of the group is the large roundworm, *Ascaris lumbricoides,* which inhabits the intestine of man, pigs and some apes. Its distribution is worldwide. Like the giant kidney worm it is very large compared with most nematodes, the full-grown females averaging 8–15in in length, with the males rather smaller and readily distinguishable because the hind end is curved in the form of a hook.

The adult worms live in the small intestine, where they feed mainly on the digested food which surrounds them, though there is some evidence that they may also bite the lining of the intestine to suck in blood and tissue juices. Unless there are very many of them, they do not do a great deal of harm.

Ripe fertilised eggs are passed into the intestine by the females and reach the exterior with the faeces. The rate of egg production is quite astonishing, each female being capable of producing more than 20 million eggs to compensate for the small chance of any egg being swallowed by a new host. Development of the eggs can only take place in the presence of plenty of oxygen and at a temperature lower than that of the host's body. They also require a certain amount of moisture. Given the right condition, though, they are capable of surviving for several years in moist soil.

The life cycle of *Ascaris* is simple, no intermediate hosts being involved. Infection takes place if any of the eggs which have developed sufficiently in the soil reach someone's mouth and are swallowed. They must, however, have reached a certain stage before they are capable of surviving and completing their development in the host's gut. When the eggs first reach the soil with the faeces, they proceed to develop until the first stage larvae have been produced. This takes about 14 days. These larvae, however, do not break out of their egg cases, but continue to develop until they have produced second stage larvae, which takes a further 7 days. This is the infective stage and no further development can take place unless the eggshells containing these larvae are swallowed.

As soon as they reach the host's intestine, the second stage larvae break out of the egg cases and the skins of the first stage larvae which are enclosing them. Then follows a complicated journey round the body. The larvae burrow into the wall of the intestine and enter a blood vessel. They can then travel round the body until they reach the liver, where they remain for a few days before re-entering the blood. Here they undergo two moults before breaking out into the air passages, travelling up the trachea to reach the back of the throat, where they are swallowed and so return to the intestine via the stomach. Now a final moult releases the young adults, which soon become mature and begin breeding. Occasionally after reaching the throat a few of them may escape being swallowed and get into the nasal passages, finally emerging down the nose—a pretty traumatic experience for the host!

Although on structural grounds the human and pig ascarids are both classified as *Ascaris lumbricoides,* there are certainly physiological differences between them, since members of the pig race cannot complete their development in man, and vice versa. Convincing proof of this was obtained by two research workers. One of them swallowed 2,000 eggs of the human strain which had reached the infective stage, while the other swallowed 500 infective eggs of the pig strain. Subsequent medical treatment resulted in more than 600 adult worms being passed by the first worker, while not a single worm was recovered from the second.

In addition to the pig and human ascarids, a number of other ascarids are found in various domestic animals. *Parascaris equorum* is a common parasite of horses, and *Toxocara cati* and *T canis* of cats and dogs respectively. Unlike most other ascarids, the life history of *Toxocara cati* may involve an intermediate host. This can be one of a number of different animals, including earthworms, mice and chickens, and it is in these intermediate hosts that the first two larval stages develop. A cat can be infected either by eating one of these infected intermediate hosts, or by directly ingesting infective eggs, in which case of course no intermediate host is involved. In the former case the second stage larvae reach the cat's intestine and remain there, undergoing a further series of moults until the adult stage is reached. But if the cat swallows infective eggs, all the larval stages must of course occur in its body, and there is a migration via the liver and lungs as there is in the case of *Ascaris lumbricoides.*

Toxocara canis is also of interest because prenatal infection often occurs, and is in fact the most common method of infection. If infective eggs are swallowed by a pregnant bitch, the larvae which hatch in her intestine migrate to her uterus, where they gain access to the developing embryos. So when the pups are born, they are already infected.

Related to the ascarids are the oxyurids, and one of these, *Enterobius vermicularis,* variously known as the pinworm, seatworm or threadworm, is a common human parasite. Unlike the majority of worm parasites it is most abundant in temperate climates, especially in Europe and North America, and is rare in tropical climates.

Enterobius is a small whitish worm, the maximum length of the females being $\frac{1}{2}$in and of the males $\frac{1}{5}$in. It lives in the large intestine and the appendix. The method used to distribute the ripe eggs is unusual. When ready to lay their eggs, the females migrate down the large intestine and pass through the anus. Once outside, they discharge their sticky eggs, which mostly adhere to the skin surrounding the anus. Migration and egg-laying usually take place at night, when the infected host is in bed. As a result some of the eggs become deposited on the bed sheets, where they soon dry out, but remain alive.

Normally after egg-laying the females return through the anus to the large intestine until a further batch of eggs has matured, but some of them fail to return. These eventually dry up and die, and then explode, releasing quantities of dried eggs which are protected by a resistant coat. Dried *Enterobius* eggs are so light that they can float about in the air.

Enterobius infection is not very serious, but when eggs are deposited around the anus, intense irritation is caused, and this results in scratching. Eggs are thus picked up, and many of them will find their way beneath fingernails. Inadequate washing of the hands can easily result in the eggs finding their way on to food, and hence into the mouth, to cause both reinfection and the infection of others who might eat the contaminated food.

Because the dried eggs are so light, they tend to be present almost everywhere in any house or institution where there are infected individuals. In such circumstances almost any sample of dust will be found to contain eggs, and because children are less

91

conscious of hygiene than adults, they are much more likely to become infected or reinfected.

The most important members of the third group of nematodes are the hookworms, which at least until recently have been among the most serious worm parasites of man. There are only two species affecting man, but they are between them extremely widespread. *Ancylostoma duodenale* occurs in the temperate areas of the northern hemisphere, where it is most common in Europe, Japan, the more northern parts of China, western Asia and parts of North Africa. *Necator americanus* is essentially the hookworm of tropical and subtropical countries, and is common throughout the southern states of America. It was in fact first discovered there, which explains why it is commonly known as the American hookworm.

Hookworms are quite small nematodes, the females approaching $\frac{1}{2}$in in length and the males somewhat smaller. On average *Ancylostoma* is a little larger and fatter than *Necator*. The adult worms live in the small intestine, where they attach themselves to the intestine wall and nourish themselves by sucking in blood and tissue fluid. To pierce the intestine wall, the mouth is provided with two chitinous plates which act as jaws. In *Ancylostoma* each plate takes the form of two well developed sharp teeth, while in *Necator* each is in the form of a blade with an extremely sharp edge.

The rate of egg production is enormous, *Necator* females producing between 5,000 and 10,000 per day and *Ancylostoma* females up to twice this number. These eggs reach the outside world via the faeces. Until they are exposed to air, they will not develop further. But once outside they proceed to develop and undergo their first two moults, feeding during this time on bacteria which develop on the faeces. The optimum temperature for rapid development is between $70°$ and $85°F$, and there must be plenty of moisture.

After the second moult, which occurs about 5 days after the eggs are deposited, the larvae have reached the infective stage, and station themselves on the highest points on the soil surface. Unlike most intestinal worms, they do not rely upon being

taken in through the mouth; instead they gain access by penetrating the skin, usually the soles of the feet. Thus people who
walk about barefooted on ground where faeces have been
deposited are most liable to become infected, though even if
shoes or sandals are worn, it is still possible to become infected
via mud kicked up on to the ankles.

Once beneath the skin the larvae enter a blood vessel and
travel round the body until they arrive in the lung capillaries,
through the walls of which they now burrow until they are free
in the air sacs. From here they work their way up the bronchi
and trachea and so reach the throat. Most of them will then be
swallowed and so reach the small intestine, where they undergo
the remainder of their moults and become adults.

Like so many worm infections, hookworms rely for their
continued existence upon the complete lack of even primitive
sanitary provisions in districts where they flourish, especially
where it is the custom for people to walk about barefooted.
Often in such communities there are special defaecation areas
situated at some distance from the village. The ground in these
areas will of course be very heavily polluted, and the consequent
rate of infection very high. Even the use of simple boards to
stand on is a partial preventive measure, and primitive latrines
are even better, for the young larvae are not capable of climbing
out of a hole in the ground. The wearing of some kind of foot-
wear, however primitive, is of course even better.

Although no animals, domestic or otherwise, can harbour or
directly transmit hookworms to man, some of them can play a
part in their dissemination. If pigs, dogs or cattle are allowed to
roam around the village, they may well eat human faeces
contaminated with hookworm eggs, which pass through their
digestive systems unharmed. These animals of course will
deposit their faeces anywhere, and so increase considerably the
areas where infective larvae can be picked up through the bare
feet.

Hookworms do not produce spectacular fatal results, as so
many other parasites do. Their effects are insidious and can
have far-reaching consequences. Because they are blood-suckers,

the total number of worms present in any infected person determines how seriously he will be affected. A hundred or so in a well nourished person will probably not even be suspected, but 500 in someone who is undernourished can be serious.

The direct effect of a hookworm infection is a steady drain on the blood, but provided the diet contains sufficient materials, especially iron, to replace the loss, the consequences of the infection will be only marginal. Thus, in judging the importance of hookworm in any population, it is not the total number of people harbouring hookworms which is important, but the number who have a large infection. A recent survey in Bengal revealed that about 80 per cent of the estimated population of 46 million were infected with hookworm, which on the face of it would seem to reveal an appalling state of affairs. In fact, however, fewer than 1 per cent of these had more than 160 worms, and almost none more than 400. So instead of being something like 35 million people urgently needing attention, only a handful required medical treatment.

Heavy infection, however, can have serious social consequences. Because of the permanent reduction in the blood, the victim suffers continuously from anaemia, which results in laziness and a lack of physical and mental energy. Historically these symptoms have played an important part in the lives of the white people living in the southern states of America. For a long time they were described as the 'poor white trash' of the south, whose laziness and lack of responsibility had brought them the poverty they deserved. It was then discovered that they were in fact the victims of large-scale hookworm infections which were stunting their physical and mental development.

In the early years of this century only about 30 per cent of families living in the country districts of the southern states possessed any kind of privy, but since the discovery of the relation between the southerners' lethargy and hookworm, great efforts have been made to persuade every family to construct one. As a result hookworm infection has fallen drastically and the people are able to work much better. Incidentally the negroes in the south become infected just as much as the whites, but for some unexplained reason hookworms have little effect upon them.

Spitting is generally regarded as a socially undesirable habit in most civilised parts of the world, but in many parts of southern Asia chewing betel-nut and spitting out the results is an extremely common habit which has probably played a significant part in keeping hookworm infection at a relatively low level wherever the habit is commonly practised. Hookworm larvae reaching the back of the throat after ascending the trachea are much more likely to be spat out along with the betel juice than swallowed. In the southern states of America in earlier times tobacco-chewing was believed to be good for the health. If it was, it was probably because the spitting got rid of hookworm larvae which would otherwise have been swallowed.

We come now to a group of nematode parasites known collectively as filariae. They are different from many other nematodes because their life histories always involve an intermediate host. Several species are important human parasites, the best known being *Wuchereria bancrofti,* which used to be known as *Filaria bancrofti.* Apart from its importance

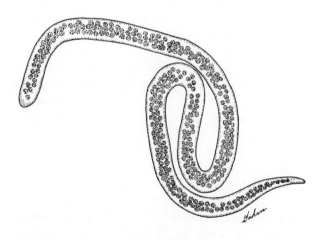

Microfilaria of the nematode *Wuchereria bancrofti*

medically, it is also of historic interest because when Manson discovered in 1878 that it was transmitted from man to man by mosquitoes acting as intermediate hosts, this was the first time that an insect, or indeed any arthropod, had been shown to be capable of transmitting a blood parasite.

Adult *Wuchereria* are extremely slender worms 3–4in long and something like $\frac{1}{100}$in in diameter—of similar dimensions, therefore, to a 3–4in length of sewing thread—the males being about half the length and diameter of the females. They live typically in the lymph glands and vessels coiled up to the extent of being inextricable, and the greatest danger to the host is that they may block the flow of lymph and cause inflammation; this blockage, as we shall see, can have most unfortunate effects.

The eggs laid by the females develop into first-stage larvae known as microfilariae, which are enclosed in transparent sheaths derived from the shells of the eggs. The behaviour of these microfilariae is curious. At times they emerge into the general blood stream, and then quite suddenly disappear as they return to the lymphatic system. Their appearance in the blood is remarkably regular. Between about 10 pm and 4 am the blood is full of them, but outside these hours it may be difficult or impossible to detect a single one in the blood.

Further development of the microfilariae is impossible in the main host's body. In order to go through the subsequent larval stages it is necessary for them to be sucked up by one of the mosquitoes which serve as intermediate hosts. These are various species of *Culex, Anopheles* and *Aedes,* and it can be no coincidence that these mosquitoes are all nocturnal in their habits, only sucking up their victims' blood at night, when there are plenty of microfilariae in the blood. During the day, when the larvae have retreated from the blood, the mosquitoes are not biting.

Once inside the mosquito's body, the microfilariae find their way to the thoracic muscles, where they remain while they proceed to grow and undergo two moults which bring them to the infective stage. These larvae now leave the thoracic muscles and work their way towards the mosquito's proboscis. While the mosquito is sucking blood from a human victim they emerge from the proboscis on to the skin, penetrating the wound when the mosquito has left so as to gain access to their main host, where the remainder of their larval development takes place. Virtually nothing is known of the subsequent development of the larvae in their human host. All that can be said is that eventually they appear as adult worms coiled up in the host's lymphatic system.

Although the adult filariae undoubtedly block the lymph vessels, their most extreme effects are probably due to the allergy they produce, causing inflammation of the lymph glands and vessels. This leads to complete blockage of the lymph system and vast accumulations of liquid in various parts of the body, leading to grotesque swellings known as elephantiasis. Various parts of the body can be affected. In severe cases the legs can swell up enormously, and an affected scrotum can expand until it weighs as much as 200lb and reaches nearly to the ground.

Wuchereria bancrofti is common and widespread in many parts of north and west Africa, Asia, the East Indian and Pacific islands, and in parts of South America. A related species, *Wuchereria malayi,* occurs throughout southern Asia from India and Ceylon to Indo-China and Malaya. Its intermediate hosts are various species of *Mansonia* and *Anopheles,* which are the dominant mosquito types in these areas.

Another interesting member of the filariae is the African eye worm *Loa loa,* of West and Central Africa. The adults are very slender worms similar in appearance to *Wuchereria* but somewhat smaller. Between the human skin and the underlying muscles there is a narrow space occupied by a loose collection of cells known as connective tissue, in which the adult worms live and wander about. Sometimes during their wanderings they may cross the surface of the eye beneath the conjunctiva, the thin layer of transparent cells which cover and protect the eye, and when they are doing this, they are of course visible. With care they can be extracted. When they settle in one place for any length of time, they set up an irritation which causes skin swellings. If the worms subsequently move away, the swellings subside, to reappear in other places where the worms have settled.

Whereas *Wuchereria* species are transmitted by night-biting mosquitoes, the intermediate hosts for *Loa* are various species of *Chrysops,* commonly known as mangrove flies. Unlike mosquitoes, these insects are on the wing and sucking blood during the daytime, not at night. Correlated with this we find that *Loa* larvae are abundant in the host's blood during the daytime, but practically disappear through the night. After they have been sucked up in a blood meal, the further development of the larvae in *Chrysops* follows a very similar course to that described for *Wuchereria* larvae in mosquitoes. In both cases, too, the infective larvae emerge from the proboscis by their own efforts and invade the final host through the wound made by the proboscis. This is in contrast to certain other insect-borne disease in which the infective larvae are forced into the host's blood with the intermediate host's saliva.

The periodicity by which *Wuchereria* larvae swarm into the blood at night when the mosquito intermediate hosts are sucking up their blood meals, or *Loa* larvae appear in the blood during the daytime when *Chrysops* is feeding, is an interesting phenomenon, and prompts the question, how do the larvae become released into the blood just at the right time of the night or the day to ensure that they will be sucked up by their appropriate intermediate hosts? Perhaps the most interesting clue is provided by the fact that if someone harbouring *Wuchereria bancrofti* flies half-way round the world, so that he now sleeps when at home he would be awake, and vice versa, the nematode larvae almost immediately adjust themselves, so that they still appear in the blood while he is sleeping, and retire when he is awake and active. So, clearly, the micro-filariae have no built-in clock mechanism. The most likely explanation of periodicity is that sleep in the host is accompanied by some subtle physiological or chemical change in the blood, and this change stimulates the microfilariae to emerge from their resting places and swarm into the blood.

The last important human nematode parasite is the guinea

worm, *Dracunculus medinensis,* distinguished by the remarkable difference in size between the two sexes. The females average 3–4ft in length but are extremely thin, having a diameter of about $\frac{1}{20}$in, while the males are usually less than 1in long. In fact they are very elusive, and not much is known about them. They were virtually unknown until 1936, when they were isolated from dogs which had been experimentally infected with *Dracunculus.* The guinea worm is widely distributed in tropical Africa and most of the southern parts of Asia, including the Near East. Like the filariae, the life history of *Dracunculus* involves an intermediate host, in this case the tiny freshwater crustacean *Cyclops.*

Human infection occurs only if water containing infected *Cyclops* is drunk. When the crustaceans reach the host's stomach, they are of course digested, and this releases any third stage infective larvae which they may contain. These now become active, penetrating the stomach wall and lying up among the tissues while the final two moults take place. When the young adult females are produced, they usually travel to the connective tissue beneath the skin, where they grow and become mature. Since the fluid in the connective tissue is a poor source of food, development is slow, and it takes a year or more before ripe eggs ready to be released are produced, and it is probably not until this stage has been reached that the victim realises that he has been infected, maybe a year or more earlier. While the females are developing, there are no symptoms to betray their presence.

Now, however, they proceed to penetrate the skin with the front ends of their bodies until they come to lie just beneath its outer layer. An irritant toxic substance is then released which causes the skin to blister immediately over the front end of each worm. Soon these blisters burst, each giving place to a shallow ulcer with a small hole in the centre. If any part of the body containing these ulcers is immersed in the cold water of a well, pond, ditch or river, the worms immediately discharge a milky fluid through the hole in the ulcer. This fluid contains hundreds of active first-stage larvae. As soon as the arm or leg is taken out of the water, the flow of fluid ceases as the aperture through which it has been passing closes up. On any subsequent immersion it is reopened and more fluid is let out. Because the ultimate release of larvae is dependent upon the part of the body containing the ulcers being immersed in cold water, the majority of the female worms coming to the surface choose those parts of the body most likely to be immersed, such as the feet, ankles and legs, and the hands and arms.

The larvae have sufficient food reserves to be able to survive in the water for several days, during which time they swim around but cannot develop further unless they are ingested by a *Cyclops.* After being swallowed, they bore through the wall of the intestine and so reach the intermediate host's body cavity. Here they undergo the usual two moults to become infective larvae. Human infection results, as already mentioned, if Cyclops containing these infective larvae are swallowed with the drinking water.

It has been estimated that as many as 48 million people may be infected with guinea worm at any one time, and suggested that the 'fiery serpents' which plagued the Israelites beside the Red Sea might have been guinea worms. The traditional method of removing guinea worms is still used, modern antiseptics making the extraction safer than it used to be. To begin the process the ulcer is immersed in cold water, which makes the worm protrude slightly from the centre of the ulcer. It is now seized and a small length wound round a piece of stick. Day by day the stick is given a turn or two and a little more of the worm withdrawn. The danger lies in pulling too hard and breaking the worm. If this happens, the portion still remaining within the ulcer can cause blood poisoning. A safe extraction takes 10–14 days.

Control of guinea worm would be very easy if people could be persuaded to change their traditional methods of drawing water. In India particularly the commonest type of well is the step well, in which, instead of the water being drawn up in buckets suspended on ropes, the people walk down the steps until they are ankle or even knee deep in the water, and then fill their buckets. This enables any guinea worms on their feet or legs to shed some of their larvae into the water, while the bucket will collect infected *Cyclops* along with the water. Infection could be further controlled if the water from the buckets was strained through muslin or similar material to remove the Cyclops before being used for drinking. But it is much easier to propose simple methods such as these than it is to get them accepted. In other places, particularly in Africa, ponds are used in the same way as step wells, people drawing water while wading into the pond instead of suspending the bucket from the bank. In many places in India where step wells have been replaced by the drawing type, the guinea worm has virtually disappeared.

Whereas the platyhelminthes and the nematodes have important parasites in nearly all their major groups, parasites feature

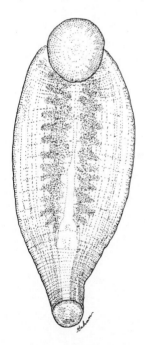

Leech, *Placobdella*

in only one of the main groups in the phylum Annelida. The phylum comprises three main classes—the earthworms or Oligochaeta, the marine worms or Polychaeta, and the leeches or Hirudinea. The first two are virtually without parasitic members, but the majority of leeches are external parasites, feeding on the blood and tissue fluids of their hosts.

Unlike other annelids, their bodies are flattened and do not have bristles. There is a sucker at each end of the body, the large hind sucker being used to anchor the body when the leech is at rest, and the smaller anterior sucker, surrounding the mouth, being used to attach the mouth to the host's body for feeding. Leeches can be divided into two groups by methods of feeding. In one group the mouth takes the form of a circular proboscis which can penetrate the host's skin, whereas in the other the mouth is armed with three sharp jaws that make a Y-shaped wound when they bite.

For most of the time leeches lead a completely free existence. Only when they need a meal do they search out an appropriate host; this does not happen very often, because they are able to take in relatively enormous amounts of blood at a single meal. The gut is exceptionally large, with lateral pouches. A single meal can suffice for several months. To prevent the blood coagulating and blocking the gut, leech saliva contains a powerful anticoagulant known as hirudin. As soon as it has sucked in sufficient blood, the leech will detach itself from its victim. Because of the hirudin the wound will often go on bleeding for a long time.

CHAPTER TEN

INTRODUCTION TO THE ARTHROPODS: LICE AND BUGS

No group of animals are more important in a study of parasitism than the arthropods, which is perhaps not surprising, for the phylum Arthropoda contains some 80 per cent of all known animal species. They are important on three counts. Not only are many arthropods parasites themselves, but many others harbour parasitic micro-organisms and act as vectors in their transmission to other animals, especially the vertebrates. Others again act as intermediate hosts for the larval stages of many helminth parasites.

The vast majority of arthropods belong to three of the subphyla into which the phylum is divided. These are the Crustacea, which contains the crabs, lobsters, prawns, shrimps and a host of smaller and more primitive members; the Arachnida, comprising the spiders, scorpions, mites and ticks; and the Insecta. In all these groups there are parasitic members as well as others which act as intermediate hosts. The insects and arachnids also have many examples which function as vectors.

The role of insects and crustaceans as intermediate hosts of various helminths has been dealt with in earlier chapters, as has the role of insects and ticks as vectors in the transmission of various pathenogenic Protozoa. But there are three other types of micro-organisms which also cause disease and some are transmitted by insects and ticks. They will be dealt with in detail in the appropriate following chapters. They are the bacteria, including the spirochaetes, the rickettsias and the viruses. Not all the members of these groups, of course, are transmitted by vectors.

Many of them are transmitted from one member of the host species to another. Of the twenty-nine orders into which the subphylum Insecta is divided only six possess members which are parasites and vectors of pathogenic organisms, but their importance cannot be exaggerated. The members of the order Mallophaga are all parasites. They are the biting or bird lice, which cause relatively little damage and are not vectors of pathogenic organisms. By contrast the closely related order Siphunculata (Anoplura), usually referred to as the true or sucking lice, are mainly parasites of mammals, and are responsible for the transmission of a number of virulent pathenogenic micro-organisms, such as those causing epidemic typhus and trench fever.

The bugs, constituting the order Hemiptera, are one of the larger insect orders, containing a great variety of members such as aphids, scale-insects, water-bugs, water-boatmen, water-scorpions, and a relatively small proportion of parasitic types, of which the bedbugs are the most important. They are extremely unpleasant creatures, but there is something rather fascinating about their ingenuity in seeking out their hosts.

Lice and bugs belong to the large division of insects known as the Exopterygota, whose life history is characterised by a series of gradual changes from the egg to the adult. After each moult the succeeding individual becomes just a little more like the final adult in appearance. These pre-adult stages are known as nymphs. In contrast, the remainder of the parasitic insects belong to the other large division, the Endopterygota. The eggs of the members of this division hatch as larvae that are quite unlike the adults in appearance. The adult stage is reached via a pupal stage, in which within a shell-like pupal case a drastic reorganisation of the larval body takes place. From this pupal case finally emerges a fully formed adult.

Three of the nine orders in this division have parasitic members. The members of the first, the Siphonaptera or fleas, are all parasites. There are many different species, certainly more than 1,000, and they are confined to the warm-blooded animals, the birds and the mammals. In themselves they are little more than irritants, but as disease vectors they are deadly, being the transmitters of two of the worst human diseases—plague and murine or endemic typhus.

The order Diptera is in terms of the number of known species the fourth largest of the twenty-nine insect orders, and its members are the true flies, characterised by the possession of only a single pair of wings, whereas all other winged insects have two pairs. There are many fly parasites and many important vectors, including the most devastating of all arthropod vectors, the mosquito. But there are many other flies which are blood-suckers, and a number of these are also disease carriers; there are yet other types of fly which are parasitic in the larval or maggot stage.

Finally among the orders which have parasitic members there is the order Hymenoptera, which has more species even than the Diptera. This is the order in which social organisation reaches its peak with the bees, wasps and ants. One section, the Parasitica, comprises the gall-wasps, ichneumon-flies and chalcid-wasps, and the section is so named because in many of the species the larvae are parasitic. Such parasites as there are in the section Aculeata, or stinging Hymenoptera, are parasites only in the adult stage.

Although the majority of the biting lice live on birds, some species are found on mammals. They all have a pair of strong pincer-like mandibles with which they nibble feathers, hair or epidermal cells. They also sometimes use these mandibles to pierce the base of growing feathers or the skin, by this means obtaining a meal of blood, though their mouthparts are not designed for actual blood-sucking.

Their whole lives are spent on their hosts, and, after mating, the females glue their eggs to feathers or hair. These hatch in a few days and usually undergo three moults during their growing period before emerging as fully mature adults. Movement from one host to another usually only occurs when two hosts come into body contact, so that one family may live on the same host for generation after generation. A few species do occasionally hitch a lift on certain flies and thus find themselves a new host. Host specificity is fairly strong, making transfer from one host species to an unrelated species unlikely.

The mouthparts of the Anoplura are highly adapted for piercing the skin of their hosts and sucking up the blood on which they feed. Almost every species of mammal has its own species of louse. Even elephants have their own louse. Unlike ticks, which take only a few enormous meals throughout their lives, sucking lice take moderate meals several times a day. In all species except the very important human louse the eggs are glued near the base of hairs. They hatch in a few days, and after three moults the adult stage is achieved after 10–20 days.

There are three different human lice—the head louse, the body louse and the crab louse. The first two are subspecies of the

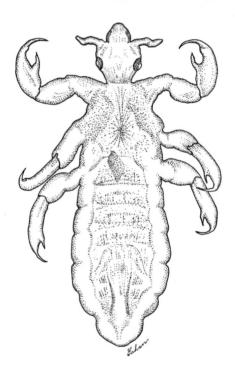

Pediculus humanus capitis (head louse)

same species, one being adapted to living on the head and the other on the body. They are *Pediculus humanus capitis,* the head louse, and *Pediculus humanus corporis,* the body louse, and there is a considerable difference in appearance between the two. It was at one time thought that they belonged to two quite separate species, but it has now been shown that if *capitis* is forced to live on the body, it will take on the characteristics of *corporis.* It has been suggested that *capitis* represents the original louse of hairy ancestral man, and that *corporis* represents an adaptation which was evolved when man lost most of his body hair. The hair which man retained in his pubic region has acquired an entirely different species, adapted, like the head louse, to living in long hair. This is the pubic or crab louse, *Phthirus pubis.*

The eggs of the head and body lice are commonly known as nits. The head louse not only lives among the hairs of the head but also it lays its eggs among them. But the body louse does not live on the body but on the clothing, reaching across to the body to feed when it requires a meal. It is possible for someone to be heavily infested with body lice but to find not one on his body when he removes his clothes. The eggs, too, are laid on the clothes.

Lice eggs are oval bodies somewhat less than 1mm in length, with a perforated lid at the larger end. Air can pass into the egg through these perforations. Female head lice lay between 80 and 100 eggs, but the body lice lay between two and three times this number. The first instar nymphs which develop in these eggs have an unusual way of escaping from them. When fully formed, they suck quantities of air into their bodies through the perforated lid, and after passing it right through the body they expel it through the anus. This results in the formation of a cushion of somewhat compressed air beyond the hind end of the body, and when the compression reaches a critical level, it is sufficient to blow off the lid of the egg shell to allow the nymph to escape. As with the Mallophaga, there are three moults before the adult stage is achieved.

Lice and other eggs are susceptible to variations in temperature and humidity. Eggs are not laid if the temperature is below 77°F, and females which are exposed to a temperature of 60°F or less for only 2–3hr out of every 24hr will only produce a fraction of their normal number of eggs. Eggs, too, will not hatch at temperatures below 70°F, and eggs exposed to temperatures below 60°F for 7 days or more will never hatch, even if subsequently warmed. This explains why in winter it is only necessary to expose clothing to the cold night air to reduce drastically or even inhibit completely the laying of body lice eggs.

Adult lice can withstand cold very well, but they are very susceptible to high temperatures, especially if the humidity is also high. The well known absence of body lice from hot countries is probably due to the high humidity between the clothes and the skin caused by profuse perspiration rather than to high temperatures as such. In Mexico, for example, they are abundant among people living on the high plateau at heights of 5,000–6,000ft above sea level, but virtually absent from the hot coastal strips. Head lice, which are less exposed to high humidity caused by sweating, are found in hotter climates than body lice.

Feeding by both body and head lice is a rather slow process. After making a puncture in the skin, they have to wait until the injected saliva has caused the host's blood capillaries to dilate through its irritation and thus produce a flow of blood which can be sucked up. This may take several minutes. Once the flow has started, the lice may take in much more blood than they can make use of. The surplus passes straight through the body and is ejected through the anus along with the faeces. Lice saliva seems to have some specificity, since human lice find more difficulty in getting blood to flow from the capillaries of other mammals than from those of man. Lice have little resistance to starvation, in contrast to ticks and bugs, which are able to survive without food for incredibly long periods. At normal temperatures they can survive for only 2–3 days, though at about 40°F they can survive for up to 10 days without food.

The effect which louse bites have on people varies considerably from person to person, and show a progression from sensitisation to immunity. Generally when a person who has never previously been bitten by lice becomes infested for the first time, the effect of the bites is initially minimal. A slight sting is felt, but there is little or no itching or redness of the skin around the area of the bites. Gradually, however, the victim acquires a sensitivity to the bites, so that after about a week they cause intense irritation and red inflamed spots. If there are a large number of lice, these effects may be accompanied by a mild general fever and a feeling of lassitude. This second phase, like the first, does not last for very long, as the sensitised skin gradually acquires immunity to the bites if they are continued.

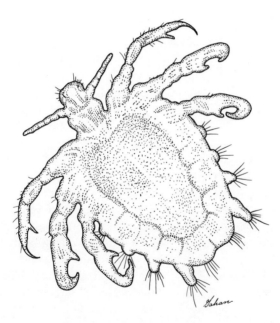

Phthirus pubis (crab louse)

This explains why people living under extremely unhygienic conditions can put up with being permanently infested with lice and apparently take no notice of them.

The sensitivity reaction to lice bites, and, as we shall see, to the bites of some other arthroped parasites, is of similar nature to an allergy, except that allergies unfortunately do not usually give way to immunity. As any sufferer from hay fever will know, continued exposure to pollen does not eventually result in a diminution and final disappearance of the hay fever symptoms.

The crab louse is quite different from the head and body louse. It owes its name to the fact that its long clawed legs give it the appearance of a tiny crab. Besides the pubic region, they are also found in other parts of the body where coarse hair grows, notably the armpits, eyebrows and beard. A particularly hairy individual may have them all over his body except the head.

The females lay only a relatively small number of eggs, two dozen or so, and these hatch in about 6 days. The young become sexually mature in another 2–3 weeks. The adults cannot survive apart from their hosts for more than about 12hr. When the adults are sucking blood, they cause intense itching. Crab lice are much less common than head and body lice, and since the most common method of transmission from one individual to another is during sexual intercourse, the French call the crab louse 'papillon d'amour'.

Lice are responsible for transmitting three important diseases —epidemic typhus, trench fever and relapsing fever. Typhus is not just one disease but several related diseases which have similar effects but are caused by different micro-organisms and transmitted by different arthropods. Epidemic typhus is caused by *Rickettsia prowazekii,* and is mainly transmitted by lice. Fleas on the other hand transmit the related *Rickettsia typhi,* which is the cause of endemic or murine typhus. Other kinds of typhus-like diseases caused by rickettsias or related species are often collectively referred to as tick typhus and are transmitted by ticks, as we shall see in Chapter 14. Trench fever is also a rickettsial disease, caused by *Rickettsia quintana.* Although, as we shall see in Chapter 14, ticks are the principal vectors for the spirochaetes belonging to the genus *Borrelia,* which cause relapsing fever, lice are also able to transmit *Borrelia recurrentis.*

Most insect vectors of micro-organisms are themselves unaffected by the parasites, which are therefore not parasites as far as they are concerned. *Rickettsia prowazekii,* however, is fatal to lice, though unfortunately it does not kill them before they have been able to transmit the organisms to man. The related typhus organisms transmitted by fleas, mites and ticks have no ill effects on their vectors, to which they are not therefore parasites.

Louse-borne typhus has been a dreaded disease in Europe for many centuries. For long periods it may be little in evidence, but whenever men are herded together in large numbers, an outbreak is likely to occur. Of course men are principally herded together in unhygienic conditions during wars. The main reason for Napoleon having to retreat from Moscow in 1812 was that louse-borne typhus was sweeping through his armies. During World War I it has been estimated that *Rickettsia prowazekii,* killed no fewer than 3 million Russians. In earlier times outbreaks of epidemic typhus were very common on ships and in prisons, in both cases involving men herded together in confined spaces. For this reason it has also been called ship fever and jail fever. To show how complicated a study of the transmission of micro-organisms can be, it is known that *Rickettsia typhi,* normally transmitted by fleas, can on occasion be transmitted by lice.

Epidemic typhus is more frequently passed around during the winter, when people are likely to be sufficiently huddled together for infected lice to be exchanged among them. Transfer of lice in these circumstances usually takes place at night, when the lice migrate from one pile of clothes to another. When it is thriving, epidemic typhus is much more deadly than endemic typhus transmitted by fleas.

Trench fever, caused by *Rickettsia quintana,* was first identified during World War I, when it caused more illness than any other disease except scabies. After the war it fell into oblivion until World War II. Where the micro-organisms went between the two wars is a mystery.

Lice are not ideal carriers of the spirochaetes that cause relapsing fever, since the majority of those ingested are killed by the lice's digestive juices. Unfortunately a few manage to penetrate the tissues of the lice and survive in their body cavities. They cannot be transmitted by bites from the lice, nor from their faeces, but are readily transmitted via the crushed bodies of the lice.

The bedbugs all belong to one family of the order Hemiptera, the Cimicidae. Their bodies are broad but extremely flattened, enabling them to hide away during the daytime in extremely narrow crevices, a favourite retiring place being behind wallpaper that has become loosened at the edges. Their blood-sucking habits are unpleasant enough, but their pungent odour is positively repulsive. This they produce from a pair of stink glands situated on the last thoracic segment, the secretion being carried along ducts which open on the upper part of the hind legs.

There are two important species of bedbugs which attack man—*Cimex lectularius,* which is widely distributed throughout temperate regions, and *Cimex hemipterus,* the Indian or tropical bedbug, which replaces it in warmer climates. There are other related species which do not normally attack man. Some of these are bird parasites, while others attack bats, and these may under certain circumstances be found in houses. If pigeons nest in the roof of a house, their particular species of bug, *Cimex columbarius,* may come down into the house. They will not normally attack the inhabitants, but they do bring their smell with them. For similar reasons *Cimex pilosellus,* a bat parasite, may also invade dwellings.

Reverse invasions can also take place, and human bedbugs can often be found in sparrow and starling nests and bat roosts in house roofs. These birds and bats can then be responsible for transferring the bugs from an already infested to an uninfested house.

Bedbugs normally come out at night, when they are most likely to find their victims in bed. When feeding, they do not usually rest on the body, preferring to cling to the host's clothing and reach across so that their piercing mouthparts can penetrate the host's flesh. This method of feeding has an important advantage. As with so many blood-sucking parasites, they tend when once attached to a victim to take in much more blood than they require. This results in the voiding of faeces containing the surplus blood. Their method of leaving their bodies among their victims' clothing means that these faeces are not voided on to the skin, where they might cause infection, but among the fibres of the clothing, where they are potentially less dangerous.

A full meal takes up to 15min, after which the bug retreats to its hiding place. To digest it completely may take a week or more, but there is evidence that after a couple of days or so the bug is ready to feed again. This is unusual, since most blood-sucking insects make no effort to obtain another meal until they have completely digested the previous one.

At the same time, in contrast to many other arthropod parasites, bedbugs can survive for long periods without food, and under experimental conditions they have been kept alive without feeding for more than a year. Although each species has its own chosen host, there is no very strict host specificity, most bugs being able to survive quite happily on the blood of some other host if necessary. The two human species have been shown to be able to thrive both on mice and chickens.

Bedbugs are famed for their ability to seek out their hosts. If the legs of a bed are placed in bowls of water to prevent the bugs from climbing up them to reach the occupants, they will climb up the wall and across the ceiling until they are over their victims, when they will loosen their hold and drop down on them. An old American verse sums up their ability :

> The lightning bug has wings of gold;
> The gold bug wings of flame;
> The bed bug has no wings at all,
> But it gets there just the same.

It seems likely that bedbugs were brought to England from the continent of Europe either by the Celts or by the Romans.

In earlier times they seem to have been tolerated as necessary evils which could even be laughed off. For example, it is recorded that in 1583 a certain Dr Penny was called to

> ... a little village called Mortlake near the Thames, to visit two noble ladies, who were much frighted by perceiving the prints of wall-lice, and were in doubt of I know not what contagion. When the matter was known, and the wall-lice were catched, he laught them out of all fear. Against those enemies of our rest in the night, our merciful God hath furnished us with remedies, that we may fetch out of old and new writers, which being used will either drive them away or kill them. For they are killed with the smoke of Oxe-dung, Horse-hair, Swallows, Scolopendra, Brimstone, Vitriol, Arsenick, Verdigrease, Lignum aloes, Bdellium Fern, Spatula foetida, Birthwort, Clematitis, Myrtils, Cummin, Lupins, Knotgrass, Gith or Cypress. But the best way is with curtains drawn about the bed, so to shut in the smoke that it can have no vent.

The materials recommended may seem bizarre, but this is probably one of the first accurate direction for fumigating a room.

Even Pepys was able to derive amusement from the attentions of bedbugs. On a trip to Bath, recorded in his diary for 12 June 1668, he found his bed good but lousy, 'which made us merry'. In the context it is clear that he was referring to bedbugs and not to lice.

Female bedbugs lay between 100 and 250 eggs each at a rate of about two a day. The nymphs undergo five moults before giving rise to adults, and before each moult the developing bug must take in a blood meal. It may in fact feed several times between one moult and the next. The rate of development depends very much upon the temperature. The stink glands of the nymphs are situated on the abdomen, not on the thorax.

The saliva of bedbugs causes irritation which can produce sensitivity and subsequently relative immunity. Excessive biting by bedbugs can result in anaemia and various nervous disorders.

114

Very severe attacks of the order of a hundred bites can produce quite severe symptoms, including heart palpitations.

Because of their wide distribution, one might expect bedbugs to be serious carriers of pathenogenic organisms, but in fact there is no firm evidence that bedbugs are responsible for the dissemination of any serious disease. Disease organisms, however, can and do quite often live in the bodies of bedbugs, but they are not able to transmit them via their saliva. Such occasional transmission as does occur is usually the result of a bedbug body carrying disease organisms being crushed on its host's body near to a puncture made by it or another bug.

The only other Hemiptera which are parasites of man are some of the Reduviidae, sometimes appropriately known as assassin bugs. These have taken to sucking vertebrate blood instead of the body juices of other insects or plant juices. They are usually brightly coloured and often large insects, most commonly found in North and South America. Most of them cause extreme pain when they penetrate the skin for a blood meal. The most important species is *Triatoma (Panstrongylus) megistus* of Brazil and other South American countries, which is one of the vectors of the protozoan Chagas' disease. Unlike the majority of its relatives, it causes no pain when it bites. Charles Darwin described this insect in his famous book, *Voyage of a Naturalist.* He calls it the benchuca, the name by which it is known in the Argentine. Under the dateline 25 March 1835 he records :

At night I experienced an attack (for it deserves no less a name) of the *Benchuca,* a species of *Reduvius,* the great black bug of the Pampas. It is most disgusting to feel soft wingless insects, about an inch long, crawling over one's body. Before sucking they are quite thin, but afterwards they become round and bloated with blood, and in this state are easily crushed. One which I caught at Iquique (for they are found in Chile and Peru) was very empty. When placed on a table, and though surrounded by people, if a finger was presented, the bold insect would immediately protrude its sucker, make a charge, and if allowed, draw blood. No pain was caused by the wound. It was curious to watch its body during the act of sucking, as in less than ten minutes it changed from being as flat as a wafer to a globular form. This one feast, for which the benchuca was indebted to one of the officers, kept it fat during four whole months; but after the first fortnight it was quite ready to have another suck.

Fortunately for the officer this particular specimen was obviously not carrying the trypanasome of Chagas' disease.

Other reduviids cause intense burning pains with their bites, and in earlier times tribal chiefs of central Asia used to keep colonies of assassin bugs for the purpose of torturing their prisoners. They would be thrown into dungeons in which the bugs were kept, and must have suffered a most agonising time. If there were no prisoners in the dungeons, the bugs would be thrown lumps of raw meat to keep them going.

CHAPTER ELEVEN

FLEAS AND BLOOD-SUCKING FLIES

The fleas, which constitute the order Siphonaptera, are a specialised group of insects believed to have evolved from the flies. They are quite a small group, containing not many more than 1,000 species, but their importance cannot be exaggerated, because they are the principal vectors for two of our most dreaded diseases—bubonic plague and endemic typhus—as well as being responsible for transmitting a number of animal diseases, including myxomatosis among rabbits.

The body of a flea is extremely compressed laterally to enable it to glide easily between the hairs or the feathers of its hosts. All fleas are parasites of either birds or mammals. The backwardly directed hairs with which the body is covered also facilitate movement over the host's body. The legs are very long and extremely powerful, giving the insect phenomenal jumping powers. The human flea, *Pulex irritans,* is able to jump something like 13in horizontally and up to nearly 8in vertically. Allowing for the difference in size, these leaps are equivalent to a man achieving a long jump of 450ft and a high jump of 275ft!

Generally speaking the mouthparts of insects are adapted for one of three different purposes. The predatory insects have powerful jaws capable of cutting up the bodies of their victims preparatory to swallowing them. Large numbers of other insects have mouthparts adapted to sucking up liquids: for example, the bees, which can take up nectar in this way, and the butterflies and moths. The third major type comprises those insects which have extremely complicated mouthparts capable of piercing the skin of their victims and then sucking up their blood;

these of course are the insects which cause most damage to man, by transmitting all kinds of pathogenic micro-organisms through their bites.

Fleas generally do not live continually on the bodies of their hosts, preferring to remain in their nests except when they need a meal. In a survey carried out in 1947 among rodent fleas the average number of fleas found in thirteen nests was more than 1,000, and only three were found on average on the bodies of each of 500 rodents examined. In the majority of species the eggs of the fleas are deposited in the burrows of the host and not on the hosts themselves. Exceptions to this rule are the cat and dog fleas, *Ctenocephalides felis* and *Ctenocephalides canis,* which lay their eggs among the hosts' hair, but even with these the eggs drop off when the animals shake themselves prior to

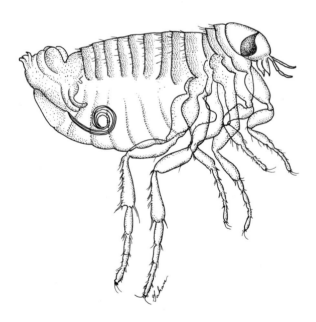

Pulex irritans (human flea—male)

going to bed. In this connection it is interesting that mammals which have no permanent nest or habitation, such as deer, antelopes, and monkeys, are virtually devoid of fleas, though they are almost invariably infested with lice.

Fleas are not so specific in their hosts as some other kinds of parasite, such as lice, though most species have one or more preferred hosts. It is this readiness to transfer temporarily which makes fleas so dangerous as carriers of disease micro-organisms from rats and mice to man. If a flea is hungry, it will suck blood from almost any warm-blooded animal which happens to be available.

Three different groups of fleas may be found in human habitations. They are the so-called human flea, *Pulex irritans,* which is believed primarily to be a parasite of the pig, but which has transferred its attentions to man; the cat and dog fleas; and rat and mouse fleas.

Unlike the lice and the bugs, fleas belong to the insect division Endopterygota, all of whose members undergo a complete metamorphosis during the life history. Flea eggs are oval pearly-white structures of relatively large size, in some species approaching one-third of the length of the females which lay them. Over a period of time, laying two or three a day, each female may produce a total of 300–400 eggs. Depending upon the temperature, these will hatch in anything between 2–3 days and a fortnight. Optimum conditions for the development of flea eggs and their subsequent larvae are high temperature and high humidity, both of which are usually found in the host nests and lairs where they are usually deposited.

The tiny larvae which hatch from the eggs are cylindrical and maggot-like, without either legs or eyes, although they are sensitive to light, avoiding it wherever possible. They feed on such organic matter as may be available to them, the most important being the faeces of adult fleas, which will usually contain a plentiful supply of their hosts' undigested blood. The duration of this larval stage will vary from about a week to several months, depending upon conditions and the species.

When eventually the time comes for them to pupate, the

larvae spin tiny cocoons which are sticky on the outside so that particles of dust adhere to them and render them inconspicuous. Unlike some insects, the adult fleas which finally emerge from these cocoons do not become sexually mature for some days, and before they will breed they must have their blood meal. For their size some fleas at least can live for a remarkably long time, up to 3 years or even more in temperate climates if they are well fed. This of course enables them to be trained for a 'flea circus'.

The susceptibility of different people to fleabites varies considerably. To some people the bite produces no reaction, the only sensation they experience being a slight tickling as the flea moves over their skin. Others, however, are extremely allergic to fleabites, finding it virtually impossible to sleep when fleas are biting them. Sensitive people can be immunised by the injection of a vaccine made from an extract of pulverised fleas, or of course by improving their standards of hygiene so that they no longer harbour fleas!

Fleas are great irritators, though it is not this that makes them such sinister insects, but the important role they play in the transmission of two extremely important and devastating diseases of man—bubonic plague and endemic or murine typhus. They not only transmit them to man from the reservoir hosts, the most important of which are rats, but are also responsible for transmitting them from one reservoir host to another. If fleas were strictly host-specific, man would not suffer from either of these dreaded diseases, because neither is carried by *Pulex irritans*. But, as already mentioned, if fleas are hungry, they are prepared to feed wherever there is a good meal available. They also act as intermediate hosts of certain tapeworms.

Historically plague has always been one of the great scourges of mankind, but only in the early years of this century was it definitely established that the particular bacterium causing it, *Pasteurella pestis,* was transferred from reservoir hosts by certain fleas, and particularly by the Oriental fat flea, *Xenopsylla cheopsis.* The method by which these fleas transmit the disease is interesting. Unlike many vectors, which are themselves unaffected by the disease they carry, the rodent fleas are such good breeding grounds for these bacteria that the latter multiply so rapidly in the fleas' digestive system that it becomes blocked with them. When such a blocked flea tries to take in a blood meal from a human victim, it is unable to swallow, and in an attempt to clear its gut of the accumulated bacteria it regurgitates them into the blood vessels of the victim from which it has been trying to obtain a meal.

The connection between plague and the Oriental rat flea was only established with final certainty in 1914 by a Plague Commission established in India in that year. Since then it has been established that *Pasteurella pestis* lives not only in rats but in many other rodents in many parts of the world, and that it can be transferred by these rodents' fleas first to the rats and then to human beings. With this knowledge it has been possible to reduce drastically the incidence of the disease in man until fewer than 1,000 cases are reported throughout the world today in any one year. Previously up to half a million cases might occur in India alone.

An outbreak usually takes the following course. Fleas from wild rodents bite rats which are living in towns or villages, and transmit the disease to them. In their turn the fleas of these town rats then transmit it to man. The stimulus to the town rat fleas to leave their real hosts and feed on man derives from the fact that *Pasteurella pestis* is fatal to its rat hosts, and just as rats are proverbially credited with leaving a dying ship, so rat fleas will leave a dying rat and seek some other host to feed on, and this may well be a human host. Fleas can still transmit plague 2 weeks after forsaking their previous host.

The first recorded instance of bubonic plague in Europe took place in 542 AD in Constantinople (Istanbul), where Gibbon estimates that up to 10,000 people died every day. By 565 AD the plague had reached Italy, where the ravages were so severe that the harvest had to be left to rot in many areas because there were not enough people to deal with it. The disorganisation caused by this great epidemic gave the *coup de grace* to the Roman Empire.

121

The next great upsurge of bubonic plague occurred in the fourteenth century. This particular epidemic is believed to have been brought into Europe from the Crimea by a group of Italian soldiers in 1347. It spread rapidly through Europe, and reached Britain in the following year, as the Black Death. Estimates of the proportion of population who lost their lives in this outbreak vary between one-quarter and three-quarters. In its wake came political, moral and religious disintegration. Men became frightened in the face of a calamity which they could not understand, and this resulted in the persecution of the Jews, which subsequent history has shown to be one of the results of mass fear.

After the Battle of Crecy (1346) the truce between England and France was prolonged because of the effects of the plague. In 1349 at least 50,000 people died in London alone, a death roll which caused the virtual suspension of business and legal work for about 2 years. Following the Black Death, the next two centuries saw periodic outbreaks of plague both in England and in other European countries. In 1656 one of the most devastating of all plagues to affect Europe began in Naples, where it is said that as many as 300,000 people died in the short space of 5 months. This particularly virulent outbreak spread northwards through Holland and Germany until it finally reached London as the Great Plague in 1665. This in fact was the last major outbreak of plague in Europe, for although periodic outbreaks occurred for another two centuries or so, these were never on the same scale as the Great Plague or the Black Death.

Bubonic plague derives its name from the fact that, besides the high fever characteristic of any disease caused by infection, it also results in large swellings, or buboes, in the lymph glands. Mortality is high, even with modern methods of treatment, so it is just as well that the disease has been almost completely conquered. In a severe local outbreak in which many people living in close proximity contract the disease it is possible for it to be transmitted directly from one person to another through the respiratory tract, and these infections result in pneumonic plague, in which the bacilli settle in the victim's lungs. This form is even more dangerous than the normal bubonic form, since the survival rate seldom exceeds 10 per cent.

The other important disease transmitted from reservoir hosts to man by flea vectors is murine or endemic typhus, the causative organism being *Rickettsia typhi.* As with plague, the principal rodent reservoir is the rat, and the principal vector the flea *Xenopsylla cheopsis.* A number of other species of flea are believed to be responsible for transmitting the rickettsias from one rodent to another, but only *cheopsis* seems implicated in transmitting them to man. Once a human population has been infected, the disease may then be transmitted from one person to another by lice. Lice, as we have already seen in the previous chapter, are killed by *Rickettsia prowazekii,* whereas fleas are unaffected by *Rickettsia typhi,* so that each infected flea can go on transmitting the disease throughout its life, and fleas, as we have already mentioned, can live for quite a time. Fortunately endemic typhus is nothing like so serious a disease in man as epidemic typhus.

Control of plague and endemic typhus poses an interesting problem. The obvious solution would seem to lie in killing off the rats; but if this is done without first killing their fleas, these will merely leave the dead rats and seek for other suitable hosts, including man, and this will accelerate the rate of human infection—hardly the most desirable outcome. So the accepted practice now is first to dust any rat burrows in the neighbourhood with DDT or some similar insecticide, which will kill off the vast majority of the rat fleas, and then a few days later administer a suitable poison to the rats.

In addition to the human, cat, dog and rat fleas which we have already mentioned, one other kind of flea is parasitic on man. This is the jigger, *Tunga penetrans,* an extremely small flea with a body length of only about 1mm. The trouble it causes, however, is out of all proportion to its size, although there is one thing that can be said in its favour—it is not known to be capable of transmitting any disease-carrying micro-organisms.

The jigger lives in sandy areas, the fertilised eggs being deposited in the soil, where they undergo normal development. The larvae feed on whatever organic debris they can find, and pupate when they have grown to full size. When the adults emerge, they mate, and it is then that they become a potential nuisance to man. The fertilised females lie in wait on the surface of the soil, ready to enter the soles of the feet of mammals, including human beings, which tread over the spot where they are waiting. In man they like to settle between the toes or under the toenails. They are now in a position to feed, and on the food they take in, which is of course the blood of their victims, the eggs are able to develop within the female's abdomen. This, swells until it achieves the size of a pea, and causes considerable distress to the host. After about a week the 100 or so eggs in the female's abdomen have become mature and are shed at the hind end of the parasite's body, which projects free from the surface of the victim's feet.

The wounds made by these enlarging females can be extremely painful, and are believed to have been responsible for the expression 'I'll be jiggered'.

The jigger is a native of tropical America, but is said to have reached Africa in 1872 aboard a vessel carrying ballast sand. In the succeeding century it has managed to establish itself over most of the continent.

We now turn to the fourth largest of the twenty-nine insect orders, the Diptera or flies. These are distinguished from all other insects in having only one pair of functional wings, the second pair having become modified as a pair of balancers or halteres. During flight these halteres vibrate at anything up to 300 times every second.

Our particular interest in the Diptera is that many of them are blood-suckers, and some of them are important transmitters of disease. Others are parasites in the larval stages, with particularly unpleasant habits. In the remainder of this chapter we shall consider the blood-suckers, with the exception of the most important of them all, the mosquito, which is so important as a disease vector that it deserves a chapter all to itself, as do also those flies with parasitic larvae (Chapters 12 and 13).

The order Diptera is divided into two suborders—the Orthorrhapha and the Cyclorrhapha. The main difference between them is that, when the latter pupate, the larval skin is retained and hardens to form an additional protective covering known as the puparium, whereas the pupae of the former are covered only with their true pupal skin. All dipteran larvae are maggot-like creatures without legs, but whereas those of the Orthorrhapha have relatively well developed heads, the Cyclorrhapha larvae have no distinct head. Blood-sucking flies are found in both suborders.

Sandflies, which belong to the suborder Orthorrhapha, probably cause even more physical irritation in tropical and subtropical climates than mosquitoes. They are tiny midge-like flies which can penetrate the finest mesh of a mosquito net. The intense irritation they cause is due to an allergic reaction. When one is bitten for the first time, the only effect felt is a slight pricking sensation, but after about a week the irritation builds up to a maximum. This is bad enough, and can cause considerable loss of efficiency, but unfortunately this minute blood-sucker is also a disease vector, transmitting the organisms responsible for sandfly fever and the various forms of leishmaniasis. Incidentally, only the females are blood-suckers.

Almost as soon as she emerges from her pupal case the female sandfly goes in search of a victim from whom she can obtain her first blood meal. She now mates, and lays her eggs in crevices in buildings or in the ground. When she is depositing her eggs, she ejects them from her abdomen with considerable force so that they can reach the innermost recesses of the crevice in which she is laying them, well beyond the reach of the tip of her abdomen. Here they are most likely to find the very high humidity without which they cannot develop. The surface of the eggs is sticky, so they adhere to the spot where they are shot.

Sandfly fever is a virus disease transmitted by *Phlebotomus papatasii,* the most common and widespread sandfly in the Old World. It is not a dangerous disease, and victims recover in 2–3

125

days. Unfortunately, however, during the short period when they are suffering from the disease, they are severely affected. In the past this has been a serious handicap during wartime, for troops sent in an emergency to areas where these sandflies are abundant have had their fighting efficiency drastically reduced just when it has been most needed.

In Chapter 6 we saw that the phylum Protozoa contained four subphyla, one of which was the Mastigophora, consisting of the flagellates, and that one of the five orders into which the subphylum was divided contained the family Trypanosomidae. This single family contains the parasitic haemoflagellates which are among the most virulent of the protozoan parasites of man. We have delayed dealing with them until now because all the important members of the family are distributed by flies. They are called haemoflagellates because they live in the blood and tissues of their victims instead of in the gut, as the intestinal flagellates do.

From the human point of view there are two important groups of haemoflagellates—*Leishmania* species, which are responsible for a number of diseases collectively known as leishmaniasis; and *Trypanosoma* species, responsible for another series of diseases referred to as trypanosomiasis.

Haemoflagellates can exist in four different morphological types or stages, each species embracing in its life history one or more, or even all, of these types. The typical trypanosome is elongated and flattened, and usually has a single flagellum projecting from the front of its body. The nucleus is situated centrally, while a second body known as the kinetoblast varies its position in the cell. The flagellum arises from the kinetoblast, so that the position of the kinetoblast determines the form of the flagellum.

In the leptomonad stage, which probably represents the primitive ancestral trypanosome form, the kinetoblast is at the extreme anterior end of the elongated cell, so that the flagellum is a simple whip-like structure projecting beyond the front end of the cell. In the crithidial stage the kinetoblast has migrated backwards to a position just in front of the central nucleus.

From the surface of the cell adjacent to it a thin undulating membrane extends forward as a normal flagellum. In the third stage, known as the trypanosome stage, the kinetoblast is now at the extreme hind end of the cell, so that the undulating membrane extends along the whole length of the body before extending forwards as a free flagellum. Sometimes the membrane is absent, and in this case the flagellum itself adheres to the body of the cell throughout its length. The last type, the leishmanial stage, is quite different from the others, for the body has become rounded and the flagellum has been lost.

Four species of *Leishmania* cause disease in man, and they all share one unfortunate characteristic. One of the most important of the body's defence systems is the large macrophage cell, whose function normally is to kill and ingest invading bacteria and other pathogenic micro-organisms. But the leishman cells are immune to the enzymes which prove fatal to other disease organisms. They can even live and multiply within the macrophages, as we shall see.

The most serious of the leishmaniases is Kala-azar, which occurs mainly in India and neighbouring countries. It is caused by *Leishmania donovani,* the most common sandfly vector being *Phlebotomus argentipes.* The life cycle of *donovani* is typical of all four species, only the effects of infection and the particular vector varying. The form in which the parasite exists in the sandfly vector is the leptomonad stage, so this is the stage which is introduced into the human victim bitten by the sandfly. But in the human blood these leptomonads are quickly transformed into rounded leishmanial forms. These now become absorbed by the macrophages in the course of their duty, but instead of being killed by them they proceed to multiply by binary fission. Although the majority of the offspring resulting from these macrophage divisions progressively invade various parts of the body, particularly the spleen, liver and bone marrow, a proportion of them will remain in the blood and be taken up by any sandfly indulging in a blood meal. Once in the body of the sandfly, the leishmania change to leptomonads, which proceed to multiply in its midgut. In 4–5 days so many have been produced that the pressure of numbers forces many of them forward into the fore part of the gut. From here they can be transferred to any human victim who may be bitten by the sandfly.

The main effects of Kala-azar are anaemia caused by the invasion of the bone marrow by the parasites, and poisoning by the products of the parasites' metabolism. Unless treated, the victim usually dies. An infantile form of Kala-azar affecting mainly children under 5 occurs around the Mediterranean, and is caused by *Leishmania infantum*. Like Kala-azar, its effects are mainly on the internal organs of the victim.

The two other forms of leishmaniasis affect the skin and surface organs of the body. *Leishmania tropica* causes skin sores and ulcers, the condition being variously known as Oriental sore, Delhi boil, Aleppo button etc, depending upon where it occurs. Although its effects are unsightly, it usually clears up without doing any serious damage. A similar infection causing ulceration of the mucus membrane of the mouth, nose and throat occurs in Central and South America, where it is usually known as Espundia. It is caused by *Leishmania braziliensis*.

For any disease caused by micro-organisms it is extremely important to know what the vectors are and what the complete life cycle of the parasite is. Not until 1942, and after many false trails had been followed, was it finally established that all species of *Leishmania* capable of parasitising man were transmitted to him by various species of sandfly. One of the difficulties in research on parasitic micro-organisms is that they are often able to live in a number of different kinds of animal but are only parasitic in some of these, causing no harm to the others. Also they may well live in a variety of insects, only one of which is capable of acting as a vector and transmitting it to man. This was the case with *Leishmania* species.

Until 1907 all that was known about leishmaniasis was that its transmission from patient to patient was not direct, and must therefore involve a vector, which would probably be an insect. But no one had ever found the parasites in any insect. In that year, however, it was discovered that *Leishmania donovani* could thrive and reproduce in the gut of bedbugs,

128

Cimex lectularius, which had been allowed to feed on Kala-azar patients. This was an unfortunate discovery, because the assumption was made that the mystery vector had now been tracked down. For the next dozen years or more all research on leishmaniasis was concentrated in trying to prove that infected bedbugs were the responsible agents in transmitting the disease, and so no other possible vector was looked for. By 1925 it had been finally proved that bedbugs were not the vectors of leishmaniasis. At about the same time leptomonads were found in certain fleas in the Mediterranean region, but these too were soon ruled out as the culprits.

Then came an observation made in 1924 by one of the scientists of the Indian Kala-azar Commission that the incidence of Kala-azar in India coincided very closely with the distribution of the sandfly *Phlebotomus argentipes,* and it was soon shown that a large percentage of these flies became infected themselves when allowed to feed on Kala-azar patients. Unfortunately all attempts to infect human volunteers by letting these infected sandflies bite them proved abortive, and it looked as though the bedbug story was to be repeated. However, research continued, and in 1939 it was discovered that if the flies were fed on raisins after their first infective blood meal, instead of allowing them to go on feeding on blood, their flagellates became so numerous that they blocked the whole of the pharynx and foregut, as we have already seen plague germs do in fleas.

Success was near. Five hamsters subjected to bites from these raisin-fed sandflies all developed leishmaniasis, and in 1942 all five volunteers developed Kala-azar on being bitten by other similarly fed sandflies. Why sandflies fed continuously on blood fail to transmit the disease while those fed on raisins after their one blood meal are able to do so is still not known, but it has been suggested that it might have something to do with an ascorbic acid deficiency in flies fed only on blood, whereas raisins will provide them with plentiful supplies of this vitamin. And it is now known that ascorbic acid is an essential growth factor for *Leishmania* species.

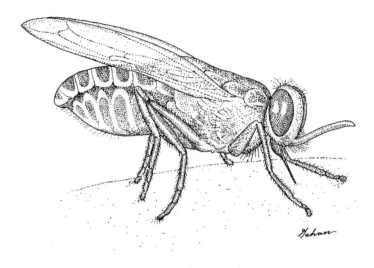

Glossina (tsetse fly)

We turn now to the tsetse flies, all of which are species of *Glossina,* and are members of the suborder Cyclorrhapha. These are the sole vectors for the trypanosomes which cause the dreaded African sleeping sickness, a much more serious disease than any of the leishmaniases.

Altogether there are about twenty species of *Glossina* which, with the exception of one species which has succeeded in penetrating as far north as some of the Arab countries, are confined to Africa and distributed from the south of the Sahara to the northern parts of South Africa. They are normally diurnal species, though one kind is said to bite on bright moonlight nights. Males as well as females are blood-suckers, but they do also suck plant juices.

Their manner of reproduction is rather unusual. They do not lay their eggs, but after internal fertilisation only one develops within the oviduct, the resulting larva being nourished by a nutritive fluid or 'milk' produced from glands in the walls of the oviduct. In the safety of the oviduct the larva grows and undergoes its full series of moults, and is only finally expelled from the female's body when it is ready to pupate. By this time it occupies practically the whole of its mother's much enlarged abdomen. As soon as one fully developed larva is born, the female gives birth to a larva every 10–12 days, this being the time taken for full development.

The larvae are deposited on loose dry soil in the shade. They then bury themselves just beneath the surface to a depth of about 1in before pupating. Between 2 and 3 weeks later, depending upon the temperature, the adult fly emerges.

The trypanosomes, which all belong to the single genus *Trypanosoma,* are parasites of vertebrates, both warm- and cold-blooded. But since they cannot be transmitted directly from one vertebrate individual to another, their life cycles involve intermediate hosts, which are always invertebrates and usually insects, with the exception of certain species which are parasites of amphibians, and these have leeches as their intermediate hosts.

The majority of trypanosomes are apparently quite harmless to their hosts, even heavy infections causing no adverse symptoms. But the best known and most important species are those which cause diseases known collectively as trypanosomiasis. One curious fact about the pathogenic trypanosomes is that they cause serious illness in some vertebrates but seem to be able to live in complete harmony with others, causing no trouble.

African sleeping sickness is caused by two different species, *Trypanosoma gambiense* and *Trypanosoma rhodesiense,* which are widespread throughout tropical Africa. The wild reservoirs for both species are the numerous game animals found in these areas, particularly antelopes, and these are not adversely affected by the trypanosomes. By contrast domestic cattle are killed by them and man seriously affected. It is therefore possible for tsetse flies and wild game to flourish in an area, even when a large proportion of the game animals are infected by trypanosomes, but where there are wild animal reservoirs and tsetse flies, it is impossible to keep domestic cattle.

Each of the two human trypanosomes has its own vector species of tsetse fly to transmit it from the wild animal reservoirs to man. *Trypanosoma gambiense* is transmitted by *Glossina palpalis* and *Glossina tachinoides,* whereas the principal vectors for *Trypanosoma rhodesiense* in different parts of Africa are *Glossina morsitans, G swynnertoni* and *G pallidipes.* Quite a number of other *Glossina* species transmit the trypanosomes from one game animal to another, but as these do not feed on human blood, they do not transmit the disease to man.

The parasites living in human blood have the typical trypanosome form, and it is in this form that they are sucked in by the tsetse flies. For 2 weeks or so they retain this form while multiplying in the gut of the fly. They then migrate to the insect's salivary glands, where they change back to a form resembling but not quite identical with typical trypanosomes, and known as metacyclical trypanosomes. This phase takes another 2 weeks or so, after which the fly becomes infective, for if it now bites a human being, these metacyclical trypanosomes pass with the saliva into the wound, and the person becomes infected. The metacyclical trypanosomes soon change to the typical trypanosome form. One surprising fact about the flies is that they seem to have quite a high degree of resistance to infection by trypanosomes, only one or two out of every 1,000 wild flies harbouring them even in areas where a very

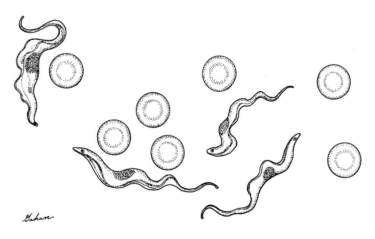

Trypanosoma vivax (haemoflagellates)

132

high proportion of all the game animals on which they feed are themselves infected.

The course taken by the infection in man is similar in the two species, but the time scale is different. For the first 2 weeks or so after infection the trypanosomes remain in the blood and the lymph glands actively multiplying, but sooner or later the parasites gain access to the cerebro-spinal fluid, and the actual state of sleeping sickness quickly follows, ending in coma and death unless the patient is treated.

Rhodesian sleeping sickness is more virulent than the Gambian form, the brain and spinal cord being invaded about a month after the initial infection. In untreated cases death results within another 2–3 months. By contrast, in the Gambian form invasion of the central nervous system may occur a few months after the initial infection, or may be delayed for as long as 7 years. During this long period it sometimes happens that the infection disappears spontaneously.

The only other important human disease caused by trypanosomes is Chagas' disease, or American trypanosomiasis caused by *Schizotrypanum cruzi*, which occurs throughout Central and South America, but especially in Mexico, Brazil and Argentina. It is not transmitted by flies but by brightly coloured reduviid bugs, which are notorious blood-suckers.

CHAPTER TWELVE

MOSQUITOES
AND PUPA-BEARING FLIES

Although, like the tsetse flies, the mosquitoes are also blood-sucking Diptera, their influence on man is so great that they deserve a chapter to themselves. Together they constitute the family Culicidae, in which there are two clearly defined sections —the anophelines, consisting of the single genus *Anopheles,* and the culicines, whose important genera are *Culex, Aedes, Culiseta* and *Mansonia.*

Of all insect pests the mosquito undoubtedly causes more distress to man than any of the others. So irritating are mosquito bites that in areas where they are really abundant they can make life almost intolerable. Unlike most insects, which tend to be more abundant in warmer than in cooler areas, mosquitoes thrive in cold climates. In Alaska, for example, they are

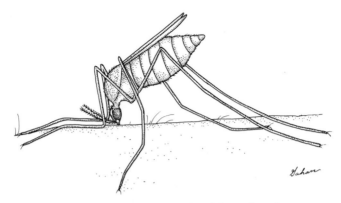

Anopheles mosquito *(Quadrimaculatus)*

present in such immense numbers that spraying equipment to deal with them is one of the principal necessities of life. But even more important than their annoyance value is their ability to act as intermediate hosts for a number of extremely serious human diseases caused by a variety of micro-organisms; most important of these diseases are malaria, yellow fever, dengue, filariasis and certain forms of encephalomyelitis. Many animal diseases also rely on mosquitoes for transmission.

There are significant differences between the anophelines and culicines which are extremely important to recognise in regions where malaria occurs, because only anophelines can transmit malaria. Adult culicines when at rest carry their bodies parallel with the surface on which they are resting, but the head is inclined downwards so that the long proboscis points towards the surface, giving the insect a hump-backed appearance. The anophelines, however, incline their whole bodies so that the head points downwards and the abdomen upwards, and the head is held in a straight line with the rest of the body.

With a few exceptions all the Culicidae breed in water, and again there are differences between the two sections. Anopheline eggs, which are oval, are laid singly, and a float of air cells around the middle enables the eggs to remain at the water surface. Culicine eggs are similarly shaped, but they lack the float; they are laid in flat masses forming a raft which floats on the surface of the water. A single raft may contain as many as 300 eggs. Although mosequito eggs will only develop in water, there are some species which lay their eggs in places which will eventually be flooded, such as marshes or water-meadows. In the late summer certain far northern species of *Aedes* lay eggs which lie dormant throughout the frozen winter, only hatching the following spring when abundant water results from the melting of the snow and ice. And in desert areas certain species lay eggs which are extremely resistant to desiccation. During their long dormant period the eggs develop to the point of hatching, so that when the rain does eventually fall, the larvae are ready to appear and complete the life cycle as quickly as possible before the water dries up again.

Anopheline and culicine larvae also show significant differences. The former lie horizontally just beneath the surface of the water, breathing by means of a pair of spiracles projecting upwards from the eighth abdominal segment to penetrate the surface. Culicine larvae, however, are only in contact with the surface of the water at their tail ends, the rest of the body hanging downwards into the water. From the hind end a pair of long breathing tubes penetrate the surface. The anopheline spiracles and the breathing tubes of the culicines are covered with an oily secretion which serves to prevent water from entering the tubes, and this is an important factor when it comes to methods of destroying mosquito larvae, as we shall see.

The differences between their positions at the water surface have an important bearing on the two larvae's methods of feeding. They both have feeding brushes on each side of the mouth, and these vibrate rapidly as they gather minute food particles and pass them into the mouth. Culicine larvae gather their food some little way below the surface, whereas the anopheline larvae turn their heads through 180° and sweep up food particles adhering to the surface film.

Larval life involves four moults, and may take from a few days to a few weeks before giving rise to the pupal stage. Insect pupae generally are unable to move, but mosquito pupae are as active as the larvae. They have something of the appearance of tadpoles, with a large head and thorax combined, and a slender active abdomen. A pair of breathing trumpets project from the upper surface of the thorax to break the surface of the water and enable the pupae to continue breathing air. The pupal stage lasts for about a week, after which the fully formed adult is ready to emerge, a process usually completed in about 5min.

The mouthparts of the mosquito are wonderfully modified to form a well nigh perfect two-way hypodermic syringe, with one tube to inject the saliva and another to suck up the treated blood. The whole piercing proboscis consists of seven separate parts, one of which, the labium or lower lip, is soft, pliable and deeply grooved, and the other six form rigid stylets which are virtually surrounded by it. Lying in the very bottom of the groove in the midline is the single hypopharynx, which has a minute channel running right through it from its tip to the salivary glands; the saliva is forced along this channel when the mosquito pierces its victim's skin. Lying on either side of the hypopharynx are the two thin maxillae, whose serrated ends enable them to play a principal part in piercing the skin. Above each maxilla is a similar thin and very sharp mandible whose tip is not, however, serrated. Above these structures is the sixth member of the piercing apparatus, the upper lip or labrum. This is so deeply grooved underneath that it virtually forms a closed tube, and it is through this tube that the blood is sucked into the mosquito's mouth.

Beneath the mouthparts there is a pair of long sensory maxillary palps. When a mosquito lands on the bare skin of a potential victim, it examines the surface with these palps in order to choose a soft spot. Having decided the exact spot where boring operations will be carried out, the mosquito uses the end of the labium to guide the stylets as they are driven into the

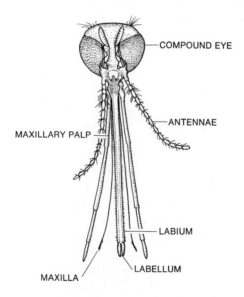

Anopheles head detail

skin. As they enter further into the skin, so the labium becomes looped out of the way. Once the puncture has been made, saliva flows into the wound through the hypopharynx, and this causes the irritation associated with the mosquito bite. This irritation probably has the effect of causing the minute blood vessels in the immediate vicinity to dilate, thus making an increased amount of blood available for the labrum to suck up.

In defence of the mosquito it must be said that it does not feed exclusively on blood, for it can get on quite well with nectar and other plant juices. Anyway, only the females can pierce the skin and suck up blood, since the male's mouth stylets are much too weak to do any piercing; male mosquitoes have to make do with such animal and plant juices as are freely available without any piercing.

The organisms which cause human malaria are four different species of *Plasmodium* belonging to the protozoan subclass Sporozoa, all of whose members are parasites. Like all Sporozoa, they have a very complicated life history, part of which takes place in the mammalian host and part in the insect vector. It is impossible for the malarial parasites to be transmitted directly from man to man or from mosquito to mosquito.

We shall start our account with an infected mosquito biting a human victim. Along with the saliva which passes down the hypopharynx there are numerous spindle-shaped bodies known as sporozoites, which represent one phase in the complicated life history of *Plasmodium*. Within about $\frac{1}{2}$hr of their injection none of these sporozoites can be detected in the blood of the human victim. They have all invaded tissue cells in various parts of the body, particularly the liver cells. They now undergo a process known as schizogony, in which the nucleus and cytoplasm of the original cell undergo repeated division until a considerable number of separate cells similar in all respects to the parent cells are produced. These second generation sporozoites leave the cells in which they are produced and enter other cells, where the process is repeated. By this time an enormous number of sporozoites will have been produced. These third generation sporozoites now leave the liver cells or other cells in which they have been produced and enter the blood. The second generation sporozoites produced in the tissue cells are technically known as cryptozoites and the following third generation as metacryptozoites. The time taken for these two cycles of a sexual reproduction to take place varies from 3 to 8 days, during which time no parasites can be detected in the blood.

The third generation sporozoites which are released into the blood now invade the red blood cells or erythrocytes. As soon as they enter the blood cells, their shape changes and they become amoeboid, moving around the cell and absorbing its cytoplasm, which it uses for its own growth. This feeding and enlarging stage is known as a trophozoite. It is characterised by the development of a clear vacuole in the middle of the cell which gives it a ring-shaped appearance. Eventually the trophozoite absorbs practically all the material of the erythrocyte, and becomes what is known as a schizont, which then undergoes schizogony, producing a number of small bodies known as merozoites. Once these have been formed, the shell of the erythrocyte bursts to release these merozoites into the blood, causing a climax in the disease, as we shall see. These merozoites now proceed to enter other erythrocytes, and the cycle is repeated.

The merozoite-trophozoite cycle may continue indefinitely, but eventually some of the merozoites switch to a different type of development. After invading the erythrocytes they change to gametocytes, which will give rise to the sexual gametes. There are two kinds of gametocytes—male microgametocytes and female macrogametocytes. These now remain dormant in the erythrocytes, unable to develop unless they are sucked up by a mosquito which has come to their host for a blood meal.

Once in the stomach of the mosquito, the gametocytes proceed to develop and produce male and female gametes. Each macrogametocyte rounds itself off and becomes a sedentary female gamete, while the microgametocytes proceed to divide a number of times, each eventually giving rise to a number of flagell-like active spermatozoa. As soon as these are released, they swim towards the female gametes and fertilise them.

The zygotes which are thus formed now make their way

between the cells of the stomach wall until they reach the outside of the stomach, and here they attach themselves as sedentary oocysts. The contents of these oocysts now proceed to divide until each becomes distended with as many as 10,000 minute sporozoites. Eventually the wall of the oocyst ruptures to release its contained sporozoites into the general body cavity of the mosquito. From here they make their way to the salivary glands, which may be packed with as many as 200,000 of them. Nothing further will happen to those sporozoites until the mosquito needs a blood meal, when enormous numbers of them will be injected into the victim of the bite, to follow the course of events already described.

To get rid of all its sporozoites the mosquito may well have to indulge in twenty or more bites, so that one infected mosquito is capable of transmitting malaria to a considerable number of people. The time lapse between the mosquito sucking up human blood containing gametocytes and the subsequent infection of another human being with these sporozoites may be several weeks, the complete development time depending upon the temperature.

The effects of malarial infection are only felt when the first merozoites are released from the erythrocytes, and this can be up to 10 days after the initial mosquito bite. When the erythrocytes burst to release their merozoites they also release into the blood toxins which cause the fever associated with malaria. As the merozoites invade other erythrocytes, so the fever subsides, to be renewed for a short time when the next generation of merozoites is released. Thus it is that malaria is an intermittent disease.

Malaria is caused mainly by three species of *Plasmodia*. Each species has a different time lapse between one generation of merozoites and the next, and thus between the bouts of fever. *Plasmodium vivax* takes just 48hr between one release of merozoites and the next, so that the fever occurs every 2 days. This form of malaria is known as tertian malaria. *Plasmodium malariae*, on the other hand, sets free a new generation of merozoites every 72hr, and is thus the cause of quartan malaria.

Plasmodium falciparum, by contrast, is less regular in its habits, liable to release merozoites into the blood at any time, and is thus responsible for irregular or quotidian fever, the pernicious malaria of many tropical countries.

Historically malaria has perhaps been more destructive as a human disease than any other. From earliest times it was recognised as a disease associated with marshes and swamps, where the air was damp, and it was assumed that the disease was a direct result of this damp air. For this reason the Italians called the disease *mala aria,* which meant literally bad air. Subsequently the name became modified to malaria. In Britain it was commonly known as ague or marsh fever.

Malaria was prevalent in Ancient Greece and Rome, and is thought to have played a significant part in the downfall of both civilisations. Until the latter half of the nineteenth century it was common in parts of England, particularly in the fenlands of Lincolnshire and Cambridgeshire, and in the coastal marshlands of East Anglia and Kent, as well as in London. In London its disappearance was due to the building of the Embankment and the draining of the swamps beside the Thames. In other areas a similar result was achieved by draining marshes and swamps in order to reclaim land for farming. But in many of these areas the mosquito species capable of transmitting malaria still exist, and the absence of malaria is simply due to the fact that there are no human sufferers from which the mosquitoes could obtain supplies of sporozoites. During World War I, however, many soldiers invalided home with malaria contracted during service overseas were sent to various parts of the East Coast to recover, and they introduced a reservoir of malaria organisms which were readily picked up by the native mosquitoes, so that for a time malaria became prevalent among the general population of these areas.

Although more than 200 species of *Anopheles* have been identified, and the majority have been shown under experimental conditions to be capable of transmitting malaria, probably less than a couple of dozen are of real importance as serious vectors of the disease. Many species live by choice well away from inhabited areas, and so are unlikely to come into contact with humans, either healthy or suffering from malaria. Others are not attracted by human blood, and will only bite if they cannot find a member of their preferred host species. Again, certain species live in inhabited areas but do not enter dwelling houses, and so seldom get the opportunity of biting a human victim.

The chief transmitters of malaria among the anophelines are the various subspecies of *Anopheles maculipennis* and certain other closely related species. The two subspecies which live in England illustrate admirably the way in which a mosquito's habits can have a profound bearing on its potential as a disease carrier. Both subspecies are, so far as is known, equally capable of harbouring and transmitting *Plasmodium* species. *Anopheles maculipennis messeae* breeds in ponds and streams, seldom attacks man and does not come into buildings. *Anopheles maculipennis atroparvus,* on the other hand, breeds in coastal salt-marsh waters and is attracted to warm, dark and ill ventilated buildings, especially where there are human beings or domestic animals. Improvements in living conditions throughout this century have reduced the number of places suitable for the members of this species, which have become increasingly rare and correspondingly less dangerous.

Until the cause of a disease is known, little can be done to combat it. The story of the discovery of the connection between malaria and the mosquito really begins in 1883, when an American doctor named King put forward the theory that malaria was transmitted by mosquitoes. At this time no one had identified the causative agent, but soon afterwards a French doctor, Laveran, isolated the protozoan parasite from a human victim. Sir Patrick Manson, a distinguished British parasitologist, accepted the mosquito idea, and at his suggestion Dr (afterwards Sir) Ronald Ross, a surgeon in the Indian Army, set out to verify the truth of Laveran's suggestion. After more than 2 years of concentrated research Ross was able in 1897 to announce that he had found the parasites in the stomach of a mosquito.

With this knowledge, and also the facts concerning the breeding of mosquitoes in water, it was at last possible to devise scientific methods of combating malaria. To eliminate adult flying insects is a well nigh impossible task, but in their larval and pupal stages mosquitoes are much more vulnerable, since they live in water. The dramatic reduction in malaria all over the world which has taken place since Ross's discovery has been achieved by two principal means. Swamps where the mosquitoes breed have been drained wherever this has been practicable, and remaining areas of water which could not be drained have been treated with applications of oil during the mosquito breeding season. The oil spreads as an extremely thin film over the surface of the water, and destroys the water-repellent properties of the anopheline spiracles. As a result water is able to enter the spiracles and the larvae are drowned. The use of mosquito nets as a further precaution has also been of great value.

If mosquitoes only transmitted malaria, they would still rank as one of man's worst enemies, but the dissemination of malaria is only one of several serious charges which can be brought against them. Yellow fever when it strikes is an even more devastating disease than malaria, though fortunately it has never been quite so widespread. The discovery of the connection between yellow fever, or yellow jack, as it was commonly called, and mosquitoes is one of the classic stories of medical research.

Yellow fever was a scourge of tropical and subtropical America, and when outbreaks occurred, there was a frightening death rate. Like malaria, it also had political consequences. Spain is said to have lost Cuba as a result of yellow fever: the Spaniards here faced almost perpetual rebellions, and whereas the rebels were more or less immune to yellow fever, it continually devastated the Spanish troops. It has also been credited with being one of the causes leading to the Spanish-American war. In 1896 the Secretary of the US State Department sent a strongly worded note to the Spanish Ambassador pointing out that Spanish-held Havana was a centre of yellow fever and a principal cause of the epidemics which at this time were ravaging the south-eastern states of the USA. Perhaps the most important result of this war was the solving of the yellow fever mystery.

As far back as 1881 Dr Carlos Finlay of Havana had suggested that yellow fever was transmitted to man by blood-sucking insects, including mosquitoes, but he was ahead of his time, and no one took the slightest notice of his suggestion. After the Spanish-American war, however, a United States Army Medical Commission was appointed to investigate yellow fever in Cuba, and if possible to ascertain the cause of the disease.

One of the leaders of the Commission, Dr Carroll, who was obviously thinking along the same lines as Dr Finlay, deliberately allowed himself to be bitten by a mosquito of the species *Aedes aegypti*. Within a few days he was seriously ill with yellow fever, and his nurse, who had had great experience in dealing with such cases, deduced that he was delirious because he said that he had contracted his fever through being bitten by a mosquito. A little later another member of the Commission, Dr Lazear, was accidentally bitten by a mosquito of the same species, and within 5 days he was dead.

The members of the Commission now decided to prove beyond doubt the guilt or innocence of *Aedes*. They erected two small huts identical in size. One of these was to be an infected hut, and into it they introduced the soiled clothing and bedding of men who had died from yellow fever. But by sealing all cracks and covering the windows and door with fine mesh netting they excluded mosquitoes. This hut was therefore completely contaminated with the disease. Three volunteers who had never had yellow fever and who could not therefore have acquired immunity to it went into this hut and stayed there for 20 days, wearing the contaminated clothing and sleeping in the contaminated bedding. At the end of their time they were all completely healthy.

The other hut, by contrast, was well ventilated, and all the bedding and the clothing worn by its inhabitants were sterilised. This hut, too, was made mosquito-proof, in this case not to keep mosquitoes out, but to keep them in. A mosquito-proof screen was arranged down the centre of the hut. In one half of the hut a volunteer who had been quarantined for 15 days to make sure that he was completely free from infection allowed himself to be bitten five times by mosquitoes which had been fed deliberately on yellow fever patients. By the fourth day he had contracted the fever. Subsequently nine other volunteers succeeded him, and each in turn contracted the disease. Meanwhile, on the other side of the dividing screen, other non-immune volunteers lived, but they were not bitten by mosquitoes. Not one of them contracted yellow fever. Thus thanks to these very brave men the mystery of yellow fever was solved, paving the way to wiping it out in many parts of the world.

One of the first results of their brave efforts and the work of Ross a little earlier on malaria was that the American army engineers were able to eliminate mosquitoes from the Panama area so that work could resume on the digging of the Panama canal, which had been halted because of the devastating effects of yellow fever and malaria on the work force. Only subsequently was it possible to show that yellow fever is caused by a virus.

Before the connection between yellow fever and *Aedes aegypti* was known, epidemics raged in tropical and subtropical cities until the majority of the population had either died or had become immune to the disease. Attempts to eliminate yellow fever have in one way succeeded better than corresponding attempts to eliminate malaria; it has been possible to develop a vaccine which gives a high degree of immunity, so that today susceptible people who are going to areas where yellow fever has not been entirely eliminated can at least be inoculated against the infection.

Unfortunately *Aedes'* habits make it much more difficult to combat than the anopheline mosquitoes responsible for the dissemination of malaria. It is almost a domestic species, seldom found more than 100yd or so from human dwellings. Its flight is silent and it hides away behind pictures or under furniture; usually the victim only knows it is about when he feels a sharp pain as the mosquito drives its stylets into his legs.

Because of its breeding habits, too, its larvae and pupae are more difficult to locate and destroy than those of the anophelines. *Aedes* shuns open stretches of water and instead lays its eggs in

145

small accumulations of rainwater held in any kind of receptacle, such as old tanks or tins, old tyres left lying around, sagging roof gutters and even abandoned coconut shells. And to treat every place where small amounts of water have accumulated would be impossible.

The position today therefore is that yellow fever has been cleared from large areas where *Aedes* is still quite common, there are other areas where yellow fever still persists but *Aedes* is virtually absent, and there are small pockets where both yellow fever and *Aedes* exist side by side. Thus in many parts of the world there are whole populations lacking any sort of immunity to yellow fever, and it is in these areas, which include India, Malaya and Australia, that the utmost vigilance is necessary to prevent the introduction of the disease. This vigilance involves two kinds of precaution. *Aedes* is widespread in all the areas mentioned, so that if someone suffering from yellow fever landed in one of these countries, he would form a reservoir of infection by which a large number of mosquitoes could become infected, and in their turn pass on the disease to other humans. Thus one patient breaking through the quarantine barrier could be the cause of an epidemic. Hence the strict immigration rules in many parts of the world designed to prevent the entry of yellow fever sufferers. There is also another way in which yellow fever could be introduced to unprotected populations : an infected mosquito from an area where yellow fever still exists might well stow away in a plane and escape when the plane landed. Even a single mosquito could easily infect a number of people, and when the disease developed in them, they would become reservoirs from which many of the native population of mosquitoes could then acquire the disease.

As with other human diseases, yellow fever is kept going in wild animals, which act as reservoirs from which humans can be reinfected. In South America various kinds of monkey are susceptible to the disease, and other mammal species are also suspected. The vectors for the transmission among these animals are certain mosquitoes belonging to the genus *Haemagogus*. Fortunately these live and breed in the forests, and do not come

146

into the villages, where they might convey the disease to the human population. Cases occur among people who work in or near the outskirts of these forests, but since *Aedes aegypti* is not found in South America, these cases are not likely to lead to epidemics.

The position in Africa is different, because here *Aedes aegypti* is a common inhabitant of most native villages. As in South America, monkeys mainly act as reservoirs for the yellow fever virus. *Aegypti* does not generally bite monkeys, and is not therefore primarily responsible for transmitting the disease to man; but a related species, *Aedes simpsoni,* feeds mainly on monkeys, transmitting yellow fever from one individual to another. It will also bite men who venture into populations near the forests in which it lives. *Simpsoni* itself will not come into the villages, but once someone has acquired the yellow fever virus from this species, that virus can then be transmitted to others by the village population of *aegypti.*

Although malaria and yellow fever are the two most deadly diseases transmitted by mosquitoes, they are also responsible for transmitting certain other diseases, notably dengue or breakbone fever. Their role in the transmission of filariasis has already been considered in Chapter 9. Dengue is caused by a virus which is believed to be related to the virus of yellow fever, but it is not so serious. It is characterised by high fever and prostration, and complete recovery takes a long time, but deaths are rare.

Dengue is widespread in warmer climates in the northern hemisphere, and often breaks out in sudden epidemics. The classic example was an outbreak which occurred in Texas in 1922, when an estimated 1 million people became infected over a period of 2–3 months. In Houston 70 per cent of the population went down with the disease.

Let us conclude this chapter with an account of an interesting group of dipteran parasites sometimes referred to collectively as the Pupipara, but comprising two separate families—the Hippoboscidae, including the deer-ked and the sheep-ked as well as a number of bird parasites; and the Nycteribiidae, a group of specialised bat parasites. The name Pupipara derives from the

147

fact that no members of the group lay eggs. At each breeding cycle the female produces a single egg which goes through all its larval stages within her oviduct, and is only shed when it is ready to pupate, which it does almost immediately after being born.

Both the pupae and the adults show marked adaptations to their parasitic mode of life. The pupae are covered with a tough leathery skin which can resist accidental crushing, and their bodies are flattened to enable them to nestle among the hair or feathers of their hosts. In the adults the legs are well developed, and end in strong claws with which they can cling to the hairs or feathers of their hosts. In contrast to some insect parasites, both sexes have mouthparts which are adapted to sucking the blood of their hosts. Some kinds have become so well adapted to their parasitic mode of life that they have lost their wings, while others shed their wings once they have become safely installed on their hosts.

Perhaps the best known of these pupiparian parasites is the sheep-ked, *Melophagus ovinus*, sometimes erroneously referred to as the sheep-tick or the sheep-louse. The adult sheep-ked is completely without wings, and even the halteres are absent. Sheep-keds look like small spiders, though of course with only three pairs of legs, as they crawl through the sheep's wool. The female ked produces only one egg at a time, which, while it is developing through all its larval stages, is nourished by secretions from the oviduct. Once it has gone through all its normal larval moults, it is born among the sheep's wool, where it almost immediately pupates, to produce a new adult within a few weeks.

The deer-ked, *Lipotena cervi*, which, as its name implies, lives on various species of deer, has a similar life history, except that when the adults emerge from the pupae, they are provided with fully developed wings, with which they are able to leave the host on which they were produced and fly away in search of another host. Having found one, they then proceed to cast off their wings, thus consigning themselves to this host for the remainder of their lives.

Another interesting species is the forest fly, *Hippobosca equina,* which lives on the wild ponies, domestic cattle and horses of the New Forest in southern England. The adults, which are winged in both sexes, are much in evidence from May to October, probably overwintering as pupae. They are very elusive, and can move rapidly through the host's fur with a crab-like movement. With their long curved claws, which enable them to take a firm grip on the hairs if danger threatens, they are not easy to dislodge. Stories are told of earlier times, when most vehicles were horse-drawn, about the effects on the town horses of a forest fly that had escaped from a visiting forest horse, which was quite used to having the flies crawling through its hair. As soon as it felt the fly crawling over its skin, the town horse often bolted, careering through the streets dragging its milk cart or delivery van behind it and scattering scared pedestrians in all directions.

In the various genera of hippoboscid bird parasites the adults still retain their wings, though these are often so reduced as to be useless for flight. Various species of *Ornithomya* are parasitic on a number of different kinds of birds, and they have fully developed functional wings. But the most interesting species are *Stenepteryx hirundinis,* which is a parasite of swallows and martins, and *Crataerina pallida,* which parasitises swifts. These species produce pupae which remain dormant in the nests throughout the winter, ready for the adults to emerge the following spring when their host birds return from their winter migrations.

No flies, however, are quite as modified as the Nycteribiidae. All the members of this family are bat parasites. The adults show no trace of wings or halteres, and they are blind. Their three pairs of legs, however, are well developed, each ending in powerful claws with which the creatures can cling to the fur of their hosts. Although they feed on the blood of bats, they seem to be completely ignored by the latter. With their powerful legs, they are able to move very quickly amid their hosts' fur. If one is disturbed when a bat is combing its fur, it will run quickly over the surface and dive down into a part that the bat has already dealt with. There are one or two of these specialised parasites on most bats, occasionally more. Compared with the size of the bat they are quite large, about $\frac{1}{5}$in long. As Dr Harrison Matthews has put it, their presence on the bat is comparable to a similar number of large shore crabs crawling about beneath our clothing.

CHAPTER THIRTEEN

INSECTS PARASITIC
IN THEIR YOUNG STAGES

We turn now to a group of insects which are in the main parasites in the larval or pupal stages and give rise to free-living adults that usually lay their eggs on the bodies of their victims. Some of these belong to the order Diptera, and others to Hymenoptera. Infestation of the body by fly maggots is known as myiasis.

Most of the important true flies or Diptera which have parasitic larvae belong to the group of warble flies and bot flies. Economically the most important species are the ox or cattle warble flies. There are a number of species, all belonging to the genus *Hypodema*. The flies themselves are large and hairy, somewhat resembling bees in appearance. Their mouthparts are rudimentary, so they are unable to feed. After emerging from the pupae they live just long enough to mate and for the females to deposit their eggs, so that the greater part of the life cycle is spent in the larval and pupal stages.

The majority of the females lay their eggs on the lower parts of the legs of the cattle, though sometimes the upper parts of the legs or even the flanks are chosen. Although these adult flies are incapable of biting or harming them in any other way, cattle seem to have an instinctive fear of warble flies, sometimes acting in a terror-stricken fashion as the flies settle on them to lay their eggs. A single fly may deposit as many as 100 eggs on one animal, each firmly attached to a hair by an adhesive substance covering the egg case.

Within about 3 days the larvae hatch and proceed to work

their way down to the base of the hair, where they penetrate the body through the skin. For the next couple of months or so they wander among the internal organs, all the time feeding and growing, finally coming to rest just beneath the skin of the back, each larva perforating the skin with a tiny hole through which it can breathe. The irritation caused by the larvae results in the surrounding tissues producing quantities of pus, thus causing cyst-like swellings known as warbles. Within the warble the larvae feeds on the pus and undergoes two moults before it becomes full grown and ready to pupate. At this stage it forces its way out through the breathing hole and falls to the ground, where it burrows beneath the surface and pupates. Complete pupation takes at least a month before the adult flies are ready to emerge.

Large infestations with warble fly larvae causes loss of condition in cattle, and the damage their hides receive runs into many millions of pounds each year; many hides are virtually ruined, becoming little or no use for making leather. Most of the damage is caused by two species, *Hypoderma bovis* and *Hypoderma lineata,* which are widespread in most parts of the world where cattle are farmed. *Hypoderma diana* is a common parasite of the various species of European deer, a related species, *Oedemangena tarandi* is a parasite of reindeer, and in India *Hypoderma crassii* attacks goats. *Hypoderma* larvae are occasionally deposited on man; they do not migrate through his body but move about just beneath the skin as migrating lumps. In Norway and Russia they sometimes invade the eye, with extremely unpleasant results.

Another unpleasant member of the group is the sheep-bot fly, *Oestrus ovis,* sometimes also known as the sheep nostril fly. Like the Pupipara we met in the previous chapter, it is viviparous, the eggs developing within the female oviduct, so that living larvae instead of eggs are deposited. They are placed by the females within the nostrils of sheep, whence they migrate to the frontal sinuses, attaching themselves to the mucous membrane by means of special mouth hooks. Here they remain for up to 9 months, feeding and growing. By spring or early summer they have reached the third larval stage, and are now having a considerable effect upon their sheep hosts, causing a dazed condition known as false gid, to distinguish it from true gid, which is caused by the invasion of the brain by the intermediate stages of a tapeworm (Chapter 8).

By this stage the larvae are fat and wrinkled, and full grown. They now release their hold upon the mucous membrane, and are sneezed out by the sheep on to the ground, where they bury themselves beneath stones or tufts of grass and change to pupae. When the adults finally emerge, they are without functional mouthparts, and are thus, like adult warble flies, able to live only long enough to mate and to distribute their larvae. Usually only a few larvae are deposited in the nostrils of any one sheep, but even a small number can cause serious loss of condition. *Oestrus ovis* was originally a European species, but it has been imported into other countries in which sheep are reared. The female flies sometimes deposit their larvae on the lips, eyes or nostrils of shepherds, presumably because they smell of sheep. Unlike *Hypoderma* larvae they seldom invade the eye itself, but young larvae about 1mm in length are sometimes found on the conjunctiva.

The deer-bot fly, *Cephenomya auribarbis,* has a similar life history to *Oestrus ovis,* and it seems that most deer become infested with it. Unlike *Oestrus,* however, it seems to have little effect upon the deer generally, though occasionally it seems to cause a kind of madness. This is thought to arise when the parasites manage to penetrate the brain, a rare occurrence.

The horse-bot fly, *Gasterophilus intestinalis,* which also attacks donkeys and mules, has a similar life history to the other bot flies. The females lay their eggs on the anterior parts of their victims' front legs. When depositing its eggs, the female hovers just above the surface of the horse's leg in an almost vertical position. The eggs are laid one at a time, and each is carefully attached to a hair. The irritation caused by the eggs stimulates the horse to lick the skin where they are attached frequently, and this repeated licking is believed to stimulate the eggs to hatch. When they do so, after about a week, the resulting larvae are carried by the tongue into the horse's mouth. Here they bury themselves in the mucous membranes and continue to develop for about a month before releasing themselves to be swallowed. As soon as they reach the stomach, they attach themselves to its wall by special mouth hooks, and here they remain for 9–10 months, by which time they have completed their larval development. They now release their hold and pass right through the gut to be released with the droppings. As soon as they reach the ground, they bury themselves and pupate, the adult flies emerging a month or two later. Since they are unable to feed, they are only able to live for 10 days at the most, but during this time they mate and the females lay their eggs.

There are several other horse-bot flies, all with broadly similar life histories. *Gasterophilus haemorrhoidalis* lays its eggs on the horse's lips, and the larvae, when they hatch, burrow into the lips or the tongue for the first few weeks of their lives, subsequently releasing themselves for later development in the stomach and duodenum. *Gasterophilus nasalis* lays its eggs on the chin and the throat, and when they hatch, the larvae crawl up into the mouth, where they settle initially between the teeth or in the gums, their irritation causing pockets of pus to collect. Again after about a month they reappear in the mouth to be swallowed, to complete their larval development in the stomach and the duodenum. In both species the larvae pass out with the droppings when they are ready to pupate, which they do after they have buried themselves in the ground.

It is an interesting fact that bot flies, like warble flies, although they inflict no pain on sheep and horses when they lay their eggs on them, nevertheless are capable of causing considerable panic, whereas other kinds of biting flies, whose bites must at times cause considerable pain, seem often to go unnoticed.

Only exceptionally do the warble flies and bot flies we have so far considered cause direct trouble to humans. But there is one related species, the human bot fly, *Dermatobia hominis,* which is a parasite of man in Central and South America. It uses an ingenious method of achieving entry for its larvae into the human body. When she is ready to lay her eggs, the female captures a large female mosquito belonging to the genus *Psoro-phora,* whose members habitually feed on human blood, and proceeds to deposit her eggs beneath the mosquito's abdomen. They are coated with a kind of quick-drying cement which ensures that they are firmly fixed. One female fly may produce as many as 200 eggs, and she will attach a dozen or more to each mosquito. In a few days the eggs are ready to hatch, and, if at this time the mosquito lands on a human being or any other warm-blooded animal to take a meal of blood, the fly larvae burst out of their egg cases and drop on the skin of the victim. As soon as the mosquito withdraws her stylets at the conclusion of her meal, the larvae effect their entry through the wound. It is said that if any larvae do not have sufficient time to emerge while the mosquito is sucking its victim's blood, they withdraw into their eggshells to await the next opportunity. Once in the skin, the larvae take 5–10 weeks to mature, reaching a length approaching 1in. A boil-like cyst similar to a cattle warble develops around the maggot, opening to the exterior by a small pore. The larva applies its hind end to this pore, through which it is thus able to breathe. These warble-like cysts can cause considerable pain. When the larvae have completed their development, they work their way out through the breathing pore and drop to the ground, where they bury themselves in the soil to pupate, a process which takes several more weeks. The complete life cycle takes 3–4 months.

The native populations of South America have various methods of removing the larvae. Tobacco juice or tobacco ash may be applied to the cyst to kill them, after which their removal is a comparatively easy matter. Another common method is to wrap a piece of animal fat tightly over the cyst; this prevents the larva from breathing, and causes it to come out of the cyst into the fat, which can then be removed, leaving the cyst empty. The best way to remove the larva is probably to use a sharp knife to enlarge the opening, and then to remove the larva carefully with forceps.

Despite its name, *Dermatobia hominis* also attacks many other animals besides man, and in South America takes the place of the warble fly as a serious pest of cattle. As well as the damage they do to hides, their presence retards growth, leading to lower meat and milk production.

The only fly which is exclusively parasitic on man is the Congo floor maggot, the larva of the fly *Auchmeromyia luteola,* which is common in all tropical parts of Africa south of the Sahara. It can only attack people who sleep on the floor, usually on mats, being unable to climb on to beds. It sucks the blood of its victims. The adult females lay their eggs on the earth floor in native huts, so that when the larvae hatch, they are in the right position to attack their victims. The eggs hatch within about 2 days of being laid, and are able within a few hours to begin sucking blood. The larval and pupal life are both quite short, and since the complete life cycle occupies only about 10 weeks, it is possible for five generations to live out their lives in 1 year.

Even elephants and rhinoceroses are attacked by bot flies and warble flies, but they of course have their own species. The stomach of a rhinoceros may harbour as many as 300 larvae of the fly *Gyrostigma,* all attached to the wall and deriving their nourishment from the food which passes through. They appear to cause little harm to the rhinos. Elephants too have a stomach bot, *Cobboldia,* while the larvae of *Ruttenia* bury themselves in the soles of their feet. But the most serious cases of myiasis in elephants is caused by a skin warble fly, *Elephantoloemus indicus,* which attacks the Indian elephant and can cause serious loss of condition among the domesticated elephants in the Indo-Malaysian area.

Another important group of flies having parasitic larvae are the members of the family *Tachinidae.* Their chosen hosts are mainly insects, especially the larvae of butterflies and moths, but they also parasitise spiders, woodlice and centipedes. There are so many of them that it is impossible to do more than outline the five main methods they use to gain access to their hosts.

In the first type the females deposit their eggs on the skin of the host larvae, leaving the larvae which hatch from their eggs to bore their way into the hosts' body. Here they feed first on the non-vital organs so that the host is not immediately killed,

156

only transferring to the organs whose destruction will result in the death of the host when they are themselves nearly ready to pupate. Perhaps the most common and best known member of the group is *Phryxe vulgaris,* a common parasite of the large white and many other butterflies and moths.

In the next group the female tachinid is able to insert her eggs into the chosen host larva. One of the best known species within this group is *Compsilura concinnata,* whose larvae have been found developing contentedly in the larvae of members of no fewer than eighteen different families.

The third method adopted to get its larvae into the chosen host's body is for the tachinid to lay its eggs on the host's favourite food plant. The host larvae feeding upon this plant also swallow the parasite eggs, which hatch in its stomach and are then able to work their way to a more suitable site for their further development.

In the last two types the eggs are laid in places likely to be frequented by suitable host larvae. In the first case the parasite larvae move about actively in search of a host, and in the second they remain stationary but rear up on their hind ends when they sense the approach of any moving creature, waving the front end of the body about hoping to make contact. Only a small minority are successful. Few of them manage to make contact, and the majority of those which do so find themselves attached to an unsuitable host. To compensate for the very considerable losses, the tachinids which adopt this method produce correspondingly more eggs.

Insects with parasitic larvae have figured prominently in the biological control of insect pests. The biochemists have worked hard to produce insecticides to kill the pests, but there are all kinds of snags in using these. They are not generally selective, and so they kill useful as well as harmful insects indiscriminately. Many of them too have proved harmful to other kinds of wild-life, and in places they have had a devastating effect on the bird population of the area. Others are not destroyed in the soil, so that over the years there can be a dangerous build-up which can cause widespread harm.

Biological control is in general much safer. The theory behind its use is that if an insect pest is causing damage to crops, somewhere in the world there probably exists another insect which will prey upon it, either by killing and eating it or by killing it by parasitising it. Once the enemy of the pest has been discovered, attempts are made to breed this enemy in large quantities and release it in the country where the affected crops are grown.

One of the earliest successful cases of biological control concerned a tachinid fly. To relate the history of this particular case will illustrate the kinds of difficulty the economic entomologist is likely to encounter in his searches. In the early years of the century the sugar planters of Hawaii were faced with the prospect of ruin, and since the growing of sugar cane was one of the island's main sources of wealth, the prospects looked bad for its economy. The pest which was attacking the canes in ever-increasing numbers was the larva or grub of a weevil. With no means of killing the weevils and their grubs, the planters resorted in desperation to hand-picking them. In one season alone about 8 million weevils were destroyed in this way, with hardly noticeable effect.

Wisely the sugar planters decided to enlist the help of an expert and in July 1906 appointed the English entomologist Frederick Muir to investigate the possibilities of combating the weevil by biological means. The successful completion of his task was to take Muir 4 years, during which he was to travel widely and suffer many disappointments. But both he and the planters were confident that sooner or later he would solve the problem.

The theory on which Muir worked, and indeed on which other economic entomologists have worked when tackling a problem of biological control, is that if an insect pest exists without any effective natural enemies, it is not native to the country where it is causing the trouble, but must at some time have been introduced from its country of origin without the simultaneous introduction of its natural enemies. Somewhere in the Indo-Pacific area, Muir argued, the original population of weevils must exist in company with their natural enemies, and this was the place he must find.

To undertake such a task must have seemed like the proverbial looking for a needle in a haystack, for the possible area of search was vast, and Muir had no clues or information to guide him. In July 1906 he set forth and for the next 6 months searched plantations and forests in various parts of southern China without sighting the weevil. Then he moved on to Malaya, where he was equally unsuccessful, and the end of his first year found him searching equally unfruitfully in Java. From here he went successively to Borneo, the Moluccas and Amboina with the same results. The first ray of hope appeared when he visited the island of Larat, because here, for the first time since leaving Hawaii, he at least found the weevil, living not on sugar cane but on sago and betel palms. For weeks, often standing in water and tormented by myriads of mosquitoes, he searched for some sign of the weevil's enemies in vain, despite opening thousands of weevil cocoons. So he concluded that Larat, like Hawaii, could not be the original home of the weevil.

But in Larat he had learned a good deal more than he already knew about the weevil, and with this additional knowledge he returned to Amboina. By this time he had already formed a theory that the natural enemy of the weevil might be one of the Tachinidae. In Amboina he found it, and it was parasitising one of the notorious weevil grubs. There was to be much disappointment, however, before the tachinids were finally released in Hawaii.

In his first attempt Muir made a collection of the tachinid flies and sent them in a cage to a collection in Hong Kong, the idea being to use Hong Kong as a half-way house where a new generation of the flies could be bred and despatched to Hawaii. Unfortunately the adult flies proved to have very short lives, and all of them were dead by the time they arrived. So Muir collected another batch and as a precaution went with them himself. But by the time he arrived these too had died.

He had learned by now that the same tachinid also existed at a place called Fak-Fak in New Guinea and, since New Guinea is much nearer to Hawaii than Amboina, he transferred his operations to Fak-Fak. Here he collected several cages of the precious flies and set sail from Port Moresby for Hawaii in high spirits, but again misfortune struck. He himself became ill with typhoid and his flies died of neglect. Having recovered, he established an intermediate breeding station in Queensland and went back to New Guinea and collected a second batch, with which he set sail from Port Moresby on 21 April 1910, nearly 4 years after his quest had begun. Despite the fact that he was again taken ill on the boat, the flies arrived safely in Queensland. From here the next stage in the final journey to Hawaii was Fiji, where a second breeding station was set up, and from the colonies successfully established here Muir was able to take the first consignment to Hawaii, landing at Honolulu on 16 August 1910. Within a year or two, by which time the tachinids had spread all over the island, the weevil was conquered. Not the least remarkable features of this story are the tenacity of Muir himself and the faith the sugar planters showed in him. Never once did they doubt that sooner or later he would succeed in solving their problem.

This in fact is not the only example of the use of a tachinid fly in controlling a pest. In the 1920s a small moth, *Levuana iridescens*, was ruining the coconut plantations in the Fiji Islands by eating all their leaves. Investigations established that the moth also existed in Java, but here it caused little harm because its larvae were parasitised by a tachinid fly. So, not without the kind of difficulty encountered earlier by Frederick Muir, Dr Tothill, the entomologist who had been made responsible for solving the problem of the defoliated palms, was at last able to transport 315 flies from Java to Fiji. The total cost per fly of this collection has been worked out at £11! After some initial difficulties in establishing the right conditions for breeding the flies, a stock of about 33,000 was built up and released to do their worst to the moths. Within a year the numbers of *Levuana iridescens* had been reduced to negligible proportions and the Fiji coconut crop had been saved.

Although practically all the successful applications of bio-

160

logical control have occurred during this century, the idea dates much further back. One of the clearest of these earlier proposals is contained in a book called *Phytologia, or the Philosophy of Agriculture and Gardening,* written by Erasmus Darwin, the father of Charles Darwin, and published in 1800. In the following extract he is dealing with aphis control.

The most ingenious manner of destroying the aphis would be effected by the propagation of its greatest enemy, the larva of the aphidivorous fly: of which I have given a print, and which is said by Reaumeur to deposit its eggs where the aphis abounds: and that, as soon as the larvae are produced, they devour hundreds round them with the necessity of no other movements but by turning to the right or left, arresting the aphis and sucking its juices. If these eggs could be collected and carefully preserved during the winter, and properly disposed on nectarine and peach trees in the early spring, or protected from injury in hot-houses, it is probable, that this plague of the aphis might be counteracted by the natural means of devouring one insect by another; as the serpent of Moses devoured those of the magicians.

We have still not finished with parasitic dipterans. One interesting group are the so-called bee flies belonging to the family Bombyliidae. They justify their common name on two counts. Many of them look like small bumble-bees, and the hosts chosen by a number of species are various kinds of solitary bees. Fabre gave a detailed description of the life history of a continental species, *Anthrax trifasciata.* It chooses as its host a certain kind of mason bee which lays its eggs singly in cells constructed in the ground and sealed after the egg has been deposited by a hard substance Fabre likened to mortar. *Anthrax* females locate these cells and lay their eggs on the ground near them. When their minute larvae hatch, they make their way towards the mortar seals looking for a tiny orifice through which they can enter the cells. These larvae are able to exist for a considerable length of time without feeding, and even when one

161

has penetrated a bee cell, it does not necessarily begin feeding at once. It waits until the bee larva is just about to pupate, and then itself undergoes a change from a slender active grub to a smooth fat grub with a small sucker-like mouth. As soon as the bee larva has pupated, this mouth is applied to its skin and its contents are slowly withdrawn to provide food for the bee fly larva. The rate at which the fly larva eats away the contents of the host's body is so regulated that just as it has sucked the bee pupa dry it is itself full grown and ready to pupate in its turn.

The fly's real problems are still ahead, however. When adult mason bees emerge from their pupal skins they are provided with powerful jaws with which they can bite their way through the mortar sealing the entrance to the cell. But adult bee flies have only weak mouthparts. They solve this problem by having not one pupal stage, which is usual with all insects that include a pupal stage in their development, but two. The first is normal, but the second develops a circle of strong spines on its head, with which it gradually manages to work its way out through the mortar. As soon as it has freed itself, the pupal case bursts open to release the adult fly.

The life histories of most other species of bee flies are similar to the one just described, but between them they choose a variety of hosts, including the larvae of mining bees and the caterpillars of butterflies and moths.

The members of a related family, the Cyrtidae or spider flies, spend at least the later part of their larval existence as internal parasites of various species of spider. Their life histories have not been so well worked out as those of the bee flies, but it seems that they lay their eggs on plants and from them hatch tiny but active larvae which go in search of their spider victims. The early larvae of some species are known to be able to jump, and presumably use this ability to jump on to a potential victim as it walks by. Once they have gained a foothold on a spider, they gain access to its body through one of its leg joints.

Despite the number of dipterans which have parasitic larvae, there is another insect order which specialises in larval parasitism to an even greater extent. This is the order Hymenoptera, the most highly developed of all the twenty-nine insect orders. Within the order two sub-orders are recognised—Symphyta, comprising the saw-flies and the wood-wasps; and Apocrita, which includes a vast assemblage comprising gall-wasps, ichneumon-flies, chalcid-wasps, bees, true wasps and ants. The sub-order Apocrita is divided into the Parasitica, which contains the gall-wasps, ichneumon-flies and chalcid wasps; and the Aculeata, which are the stinging Hymenoptera and include the bees, wasps and ants. These, too have evolved a highly developed social system. The Parasitica are so-called because in the vast majority of species the larvae live as parasites on other animals, usually insects. But the Aculeata also contain many members whose larvae lead a parasitic existence.

Of the Parasitica only the ichneumon-flies and the chalcid-wasps are animal parasites. Two divisions of ichneumon-flies are recognised, the true ichneumon-flies belonging to the family Ichneumonidae and the so-called supplementary ichneumon-

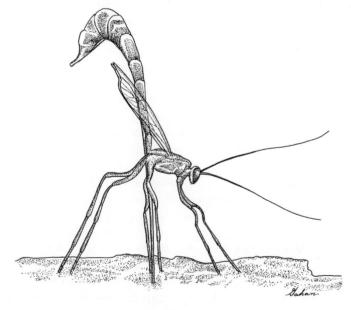

Ichneumon-fly, *Megarrhyssa macrura*

flies constituting the family Braconidae. Chalcid-wasps are extremely small insects, but are just as efficient at laying their eggs on the larvae of other insects as are the ichneumon-flies. Some of them have played extremely important roles in biological control.

One of the best known ichneumon parasites is *Apanteles glomeratus,* a small braconid which specialises in parasitising the large cabbage white butterfly, *Pieris brassicae,* one of the major pests of vegetable gardeners. Like all ichneumon-flies and chalcid-wasps, *Apanteles* has a well developed ovipositor with which it pierces the skin of butterfly caterpillars, and, because it is so small compared with its victim, it may lay as many as 100 eggs on a single host caterpillar. When the eggs hatch, they carefully avoid feeding on any of the vital organs of the caterpillar, which might kill it, but concentrate on the stores of fat that are really meant to provide the future butterfly pupa with the material it needs for the transformation from caterpillar to butterfly. By the time the caterpillar has reached the stage at which it would normally pupate, it dies, owing to the loss of these fat stores. By this time too the ichneumon larvae are ready to pupate. This they do by emerging from the body of the dead caterpillar and forming tiny yellow cocoons on its surface, from which adult ichneumons eventually appear.

This, however, is only one small part of the saga of the parasite of the large white butterfly. *Apanteles* too has its parasitic enemies, even when it is apparently safely hidden away inside the body of a *Pieris* caterpillar. A much smaller ichneumon, *Hemiteles nannus,* is able to locate the *Apanteles* larvae with its ovipositor and lay its eggs in them; *Hemiteles* is therefore a hyperparasite, that is a parasite of a parasite. In this case what finally emerges from the original *Pieris* caterpillar is not a butterfly, nor a number of *Apanteles,* but a veritable swarm of minute adult *Hemiteles.* Incredibly, even this may not be the end of this complicated story, for certain chalcid-wasps even smaller than *Hemiteles* may deposit their eggs in the *Hemiteles* larvae, which are inside the *Apanteles* larvae, which are themselves inside the *Pieris* larva!

To add to the discomfort of *Pieris,* another chalcid-wasp, *Pteromalus puparum,* which ignores *Pieris* caterpillars, has been seen to lie in wait beside one until it has pupated, and then lay its eggs in the resulting pupa, which will of course be destroyed nourishing the chalcid larvae. Bearing in mind that *Pieris* caterpillars also have to contend with the depredations of birds, and that they are prone to various diseases, it is perhaps not surprising that out of every 10,000 young caterpillars which hatch, calculations suggest only about thirty survive to give rise in their turn to adult butterflies. It also seems certain that if such a toll were not taken, it would be virtually imposible to cultivate cabbages and other brassicas in countries where the butterfly flourished.

Certain tiny braconids belonging to the genera *Aphidius* and *Praon* lay their eggs in aphids, and are therefore of great benefit wherever aphids occur in numbers—which means virtually everywhere! If they could be bred and distributed, they might well be valuable allies in biological control.

One of the largest and economically most useful of all the ichneumon-flies is *Rhyssa persuasoria.* It is an impressive insect with a bluish-black body about 1¼in long. Its abdomen carries white markings, and its legs are red. The most astonishing feature is the female's ovipositor, which measures 1½in in length, and is an extremely strong boring tool, as we shall see.

Rhyssa females lay their eggs on the larvae of the giant wood-

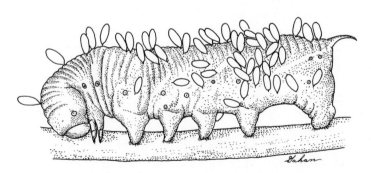

Sphinx moth caterpillar parasitised by Braconid wasp larvae

wasp, *Sirex gigas,* itself an impressive insect. It is yellow and black and about $1\frac{1}{2}$in long, with a stout ovipositor nearly as long as its body. It is often feared because of its superficial resemblance to a hornet, but it is completely harmless. It is nevertheless a pest, because it lays its eggs in the trunks of pine trees and other conifers, the female being able to bore a hole more than 1in deep into solid wood, and then deposit a single egg at the bottom. When the larvae hatch, they proceed to bore their way through the wood, feeding on the shavings scraped off by their powerful jaws. For 2–3 years they continue their meanderings, which may take them deep into the trunk, but when they are at last ready to pupate, they always come to within $\frac{1}{2}$in of the surface, so that when the adult finally emerges, it has only this thickness of wood to negotiate before it breaks free. Obviously a heavy infestation of wood-wasp larvae can considerably reduce the value of a tree when it is felled for timber.

But thanks to *Rhyssa,* the giant wood-wasp does not have things all its own way. As soon as its larva begins to develop and before it has had time to move deeper into the trunk, a female *Rhyssa* is liable to come along, and with her long ovipositor bore a hole through the wood until she has located the larva, on which she then proceeds to deposit an egg. The most interesting aspect of her behaviour is the way in which she can locate her own larva victim; ichneumons are known to possess a strong chemical or smell sense, which the female clearly uses to guide her boring activities. And she is able to achieve her objective remarkably quickly. Under observation a *Rhyssa* female has been seen to bore a hole $1\frac{1}{4}$in deep into solid wood in 20min. With the effort involved it is just as well from her point of view that she seldom fails, because of course she has many more than one egg to dispose of in this way. The channel through her ovipositor is so narrow that the egg is squeezed completely out of shape as it travels down, but resumes its normal shape when it is deposited on the host's body.

Rhyssa has become one of the classic examples of biological control. In the 1920s the New Zealand Government was becoming very concerned about the increasing ravages of the wood-wasp *Sirex noctilio,* which was causing widespread damage to coniferous trees and rendering the timber worthless. There were no natural enemies of this particular species of wood-wasp in New Zealand. Accordingly it was arranged for specimens of *Rhyssa persuasoria* to be collected in England and shipped out to New Zealand in the hope that, once established, they would be able to control the ravages of the New Zealand wood-wasp. This turned out to be one of the most spectacular examples of the control of a pest by biological methods, for by 1936 the wood-wasp had been brought under control.

One of the major steps in the establishment of biological control as an economic science was made in 1927 when the Imperial Institute of Entomology opened the Farnham House Laboratory at Farnham Royal in Buckinghamshire. The function of this laboratory was to be the collection and breeding of insects known to be parasites of others that were known to be economically undesirable, and the collection of the *Rhyssa* larvae for transmission to New Zealand was one of the laboratory's first notable successes. Some idea of the importance of the Farnham Laboratory can be gained from the fact that in the first 12 years of its existence up to the outbreak of World War II about 40 million insects or their larvae or pupae had been distributed to almost every part of the world, and had achieved notable successes in combating various insect pests.

Like the ichneumon-flies, the chalcid-wasps constitute another large group of parasites. Very few members of either group parasitise adult insects, but whereas ichneumon-flies specialise in larval hosts, various kinds of chalcid-wasps lay their eggs on or in pupae, larvae or even eggs, and a considerable number of them are hyperparasites. One group, the *Mymaridae* or fairy-flies, are among the smallest of all insects, with adults scarcely achieving $\frac{1}{50}$in in length, and it is mainly these which choose the eggs of other insects, especially moths, as the hosts for their own larvae.

Some of these minute chalcids have developed an extra-ordinary method of multiplying. Usually a single egg is laid in

167

each moth egg, and when the moth larva hatches, the chalcid egg is developing inside its body in the form of an elongated mass of cells. But instead of forming themselves into a single chalcid larva, the mass breaks up into a number of separate groups of cells, each group then going on to form a separate larva, so that from a single egg a number of larvae are formed, a phenomenon known as polyembryony. Depending upon the species, the number of larvae arising from a single egg may vary between two and more than fifty, occasionally many more. The record seems to be held by *Litomastix,* which lays its eggs inside those of the silver Y moth, *Plusia gamma.* From a single egg as many as 1,000 parasitic larvae may develop. Exceptionally the chalcid may lay three eggs in the same moth egg, which thus produces upward of 3,000 chalcid larvae.

Because they are such a successful group of parasites, it is not surprising that some species of chalcid-wasps have featured in a number of successful experiments in biological control. One of the major greenhouse pests used to be the greenhouse white fly, *Trialeurodes vaporariorum,* which is not a fly at all but one of the large group of plant-sucking bugs that includes the aphids and the scale-insects. Until a method of controlling them by biological means was developed, the only effective way to control the white fly was to fumigate greenhouses with hydrogen cyanide, an extremely poisonous gas whose use resulted in a number of deaths. Growers of tomatoes and cucumbers suffered particularly. The white fly was unknown in Britain until the early years of this century, when presumably it was imported with plants brought into the country. There is some evidence that it came from Brazil.

The discovery of an efficient method of controlling white fly biologically was accidental. In July 1926 a number of minute chalcid-wasps about $\frac{1}{40}$in long were discovered laying their eggs on the immature stages of the white fly in a small greenhouse at Elstree in Hertfordshire. The discoverer immediately contacted the Cheshunt Experimental Station, and by September E. R. Speyer, one of the Station's entomologists, had already succeeded in rearing initial stocks of the chalcid, which were identified as *Encarsia formosa,* a species not recorded in Britain before. Presumably it had been introduced recently with imported plants, though from where no one knew.

By the following summer it was already clear that *Encarsia* was capable of clearing white fly completely from greenhouses, and large stocks of the chalcid were already being distributed all over the country. Subsequently stocks of the valuable parasite were sent to many parts of the world where *Trialeurodes* existed as a greenhouse pest.

The white fly and the aphids, members of the order *Hemiptera* or bugs, belong to the large division of insects which do not have distinct larval and pupal stages in their life histories, but develop gradually through a series of moults from the first nymphal stage, which hatches from the egg, to the final adult, which emerges when the last nymphal stage undergoes its inevitable moult. Thus, unlike larvae, nymphs do bear a certain structural resemblance to the final adults. *Encarsia* females lay their eggs upon the various nymphal stages of *Trialeurodes,* but not on the adults. As with so many of their relatives, including the aphids, parthenogenesis is much more common in white flies than sexual reproduction, the females producing eggs which develop without fertilisation, and this enables them to multiply very rapidly. Males are very rarely found.

Fortunately for those engaged in the search for parasitised white flies the nymphs on which the chalcid-wasps have laid their eggs can be readily identified because they turn black. This provides a simple means of distributing the chalcid-wasp eggs. Tomato leaves bearing the blackened nymphs are collected from infected plants and sent to wherever the parasites are required. When they reach their destination, they are simply hung in any greenhouse infested with white fly. Eventually the chalcid eggs hatch within the parasitised nymphs to produce larvae, and these in turn will develop through a series of moults to the pupal stage, from which finally the adult chalcids will emerge to begin their valuable work of destroying the white flies.

Another chalcid, *Aphelinus mali,* is very effective at controlling the woolly aphis, *Eriosoma langerum,* popularly known as American blight. It is a typical plant-sucking aphis, but it has a remarkable ability to hide itself under a white waxy covering which looks like cotton wool, and which is formed from glands in its skin. The wool is of course conspicuous, but few enemies will risk getting themselves 'gummed up' in an attempt to locate the insect which is hiding underneath. American blight is a pest of apple trees in particular, and its effects can be very serious in apple-growing areas in many parts of the world. *Aphelinus mali* was first discovered in America doing noble work by probing the wool with its long ovipositor and depositing its eggs in the bodies of the aphids when it had located them. The effects of woolly aphis were particularly severe in New Zealand, but after its discovery large numbers of *Aphelinus* were collected and sent there. Within a few years the chalcids had reduced the woolly aphids to manageable proportions. Introductions to South America, Australia and Tasmania had similar beneficial results. Attempts to establish *Aphelinus* in Britain, however, have not been successful. Presumably the climate is unfavourable, for after a time introduced colonies die out. To be used effectively to control the aphids, captive stocks would have to be maintained and batches released periodically in infested orchards.

A scale-insect has also been controlled biologically by a chalcid-wasp. It is the British brown scale, *Coccus coryli,* which lives on various trees, especially the hawthorn. It is not really a pest, since it is kept under control by the chalcid *Blastothrix sericea.* But in 1903 it suddenly appeared in British Columbia, probably having come over from Britain with various ornamental trees imported to provide shade in Vancouver. Unfortunately its enemy *Blastothrix* was not imported with it, and so it was able to flourish and spread all along the coastal region of British Columbia. The effects of a heavy infestation are to cause the leaves of the trees to fall prematurely. The Vancouver authorities had to resort to frequent oil spraying of their maples to arrest this leaf fall. In 1928, however, a programme of biological control was launched. Large numbers of adult chalcids were collected. The problem of keeping them alive while they crossed the Atlantic before the days of air travel was solved by putting them in cold storage in glass tubes, with split raisins to provide food. From this initial stock large populations were propagated and finally released. Success was rapid, and within 4 years *Coccus coryli* had been almost wiped out.

Our last example of a chalcid parasite is *Trichogramma evanescens*. Whereas most parasites show host specificity, only parasitising one or at the most a few closely related species, *Trichogramma* is known to lay its eggs in the eggs of at least 150 different insect species belonging to no fewer than seven of the twenty-nine different insect orders. The majority of these hosts are, however, various species of moth, and it is as a moth parasite that it has proved most valuable. Unlike other chalcid females, the female *Trichogramma* examines a potential host egg very carefully, walking all over it, before inserting her ovipositor to deposit a single egg. This has the effect of leaving her scent on the parasitised egg, and this can be detected by any subsequent chalcid female that may examine the egg, and thus dissuade her from laying one of her eggs in a host egg which is already occupied.

In America *Trichogramma* has been used to combat the gypsy moth in areas where it is a pest. In order to build up stocks of *Trichogramma* for distribution many moth eggs are exposed to the chalcid females. Not all of them will be parasitised, but from the point of view of biological control it is important to be able to distribute eggs in which parasite eggs have been laid, and to accomplish this separation of the two types of gypsy moth eggs a most ingenious method was developed. Gypsy moth eggs are covered with a thick pad of hairs, so that they tend to stick together in masses. To make their separation easy a machine was devised to remove the hairs by rubbing the eggs gently between two discs, one moving and one stationary. Having removed the hairs and thus ensured that the eggs would not adhere to one another, the researchers needed a device to separate those eggs in which a *Trichogramma* egg had been laid from those in which it had not. Their solution was a chute down which the eggs were dropped, to land on a metal plate from which they bounced off. The eggs which were not parasitised bounced better than those which were, and the machine was so constructed that the better non-parasitised bouncers landed in a separate compartment from those which were parasitised, thus ensuring that batches of eggs sent out to combat the gypsy moth would each eventually produce a chalcid which would contribute to the destruction of the moth.

Although the organisms concerned are not Hymenoptera, this seems to be an appropriate place to consider one or two other outstanding examples of biological control using insects. Perhaps the most astonishing story of all is the conquest of the prickly pear in Australia. Prickly pears are in fact a number of different but related species of cactus native to Mexico and other parts of Central and South America. The reason for their original, and ultimately disastrous, introduction to Australia was economic. One of the many insects which live and feed upon prickly pears is the cochineal insect, *Dactylopus coccus,* a scale-insect whose value lies in the fact that its dried body may be ground up to produce the cochineal that is used as a colouring matter for food and as a dye. In an attempt to establish a cochineal industry in Australia a number of prickly pears were introduced from Central America in 1787. In those days, of course, the potential dangers of introducing alien plants or animals to new countries was not appreciated. Altogether quite a number of species of prickly pear became established, but the most important, because they became major pests, were two species, *Opuntia inermis* and *Opuntia stricta.* The speed with which they spread was remarkable. By 1925 it was calculated that the area occupied by these two species, and therefore useless for any other purpose, in Queensland and New South Wales alone exceeded 60 million acres. Physical clearance of this land was quite impossible. The cost would have been prohibitive, and the value of the land when cleared probably not more than £1 per acre.

The Australian Government, alarmed by the immensity of the problem, appealed to the scientists for help. The first step was to send an expedition to Mexico and other parts of Central and South America where the cacti grew in order to find out whether they had any insect enemies which might be introduced into Australia to control them. The scientists sent were cautious. They had already seen how disastrous introducing an animal or plant into a land to which it was not native could be. Both the prickly pear and the rabbit were introduced, and both had become major pests. So although they might find an apparently efficient enemy of the cacti, they would not risk importing it into Australia until they were completely satisfied that such an introduction would not in time become yet another pest.

Something like 150 different insects found feeding on the cacti were considered, but in the end it was a small moth from the Argentine, *Cactoblastis cactorum,* which finally solved the problem. The eggs are laid on the surface of the cactus, and as soon as the caterpillars hatch, they burrow into the succulent tissues in great companies which soon reduce the whole plant to a mass of dead rotting pulp. Initially in 1925 only 2,750 eggs were brought over from the Argentine, but as soon as it became clear that the moth was indeed a highly successful parasite of the cacti, large numbers were bred for distribution to areas where the cacti were the greatest trouble. To give some idea of the scale of this distribution in the three years from 1928 to 1930, some 3,000 million eggs were distributed. But the attempt to control and destroy the cacti by biological means was a phenomenal success. Within a few years well over 20 million acres in Queensland alone had been cleared and developed, an area incidentally equal to about half the total area of England.

Another group of insects which have proved their worth in biological control are the ladybirds, both adults and larvae. These are colourful beetles belonging to the family *Coccinellidae.* The larvae in particular are carnivorous, feeding voraciously on aphids and scale-insects. The adult females lay their eggs in small batches wherever they can find a flourishing colony of these plant bugs, so that when the larvae hatch they have an abundant supply of living food immediately available.

In 1872 citrus fruit growing was becoming well established in California, but in that year a species of scale-insect known as the cottony-cushion scale-insect, *Icerya purchasi,* appeared on the trees. It was recognised as belonging to the same species as one commonly found on citrus trees in Australia, whence the new Californian arrival is believed to have been accidentally introduced. During the next 15 years the spread of the insect was rapid and the industry was faced with a crisis. It was known, however, that although the Australian citrus trees were also attacked by the same scale-insect, the damage it caused was not serious. Accordingly the United States Department of Agriculture sent one of its scientists, Albert Koebele, to Australia in the hope that he might be able to discover an enemy of the scale-insect which might be introduced to California.

What he found was that the Australian scale-insects were kept within bounds by a bright red ladybird, *Vedalia cardinalis.* At once he collected some adults and despatched them to America. Only 129 were still alive when the package arrived at its destination, but on their release they settled down at once to feed on the scale-insects and to reproduce. They flourished, and within a couple of years subsidiary stocks had been distributed to virtually every orange and lemon grove in California. Within weeks of their arrival on any plantation the ladybirds were already reducing the scale-insects to manageable proportions. The saving to the citrus fruit growers must have amounted to millions of dollars, yet the total cost of obtaining the first 129 ladybirds was a mere $1,500—a sound investment indeed!

During the 1920s the coconut scale-insect, *Aspidiotus destructor,* was increasing rapidly among the coconut palms in Fiji and becoming a serious threat to the coconut industry. The same scale-insect was known to exist in Trinidad, but here it caused little trouble. The reason was found to be that it was kept in bounds by the ladybird species *Cryptognatha nodiceps.* Batches of the ladybirds were accordingly collected in Trinidad and released in Fiji in 1928 and they were immediately successful. By the end of 1929 the pest had been brought completely under control.

Among the Aculeata, or stinging Hymenoptera, there are two groups of parasites—the solitary wasps and the cuckoo bees. Bees feed entirely on plant products, mainly nectar and pollen, whereas the larvae of all wasps are carnivorous. The solitary wasps have the most remarkable methods of ensuring that their larvae, when they hatch, have a plentiful supply of animal food readily available for them. Although there are many different species and individual differences in the way in which each species provides for its larvae, they all follow essentially similar patterns of behaviour. The female lays her eggs singly, and before doing so prepares some kind of chamber or cell where both the egg and the larva which hatches from it can live in comparative safety. The potter wasps construct cells made of earth which they attach to the stems of low grow-ing shrubs, whereas the digger wasps excavate a hole in the ground and convert this into a cell suitable for the reception of an egg complete with the food supply needed by the subsequent larva.

Having constructed the cell, the female wasp then flies away to capture the necessary food store, which may be an adult insect, an insect larva or nymph, or a spider. She stings her prey, the venom from her sting paralysing her victim perm-anently, not usually killing it. It is then taken back to the pre-pared cell, where a single egg is laid on it. The entrance to the cell is then sealed. When the egg hatches, the tiny larva has sufficient food for its complete growth up to the time it is ready to pupate. The virtue of the mother's sting merely paralysing the victim is that it will not begin to decay while the wasp larva is feeding on it. In those species in which the victim is killed by the sting of the mother wasp, her venom acts also as an antiseptic, which ensures that the victim will not decay before it is consumed.

Two kinds of wasps which have solitary habits really belong to the group of social or true wasps, though their nesting habits are different from those of the true social wasps. The best known of these social wasps turned solitary is *Eumenes coarctata*, the potter wasp, which constructs the suspended earth cell already mentioned. Once her cell is ready, the female wasp goes out collecting caterpillars, and when she has practically filled the

175

cell with a number of them, she proceeds to lay a single egg, not on the top of the pile of caterpillars but suspended from the roof of the cell on the end of a thread. When the tiny larva hatches, it remains suspended while it feeds on the caterpillar on top of the pile. By this time it will have grown sufficiently to be in no danger of being crushed by the pile of caterpillars, which it might have been when it was first hatched. *Odynerus* species have similar habits, but some of them pack their egg cells with beetle larvae instead of the caterpillars of butterflies and moths.

The majority of true solitary wasps which excavate egg burrows show no further interest in their potential offspring once they have excavated the cell, provisioned it with the insect of their choice, and laid the single egg. But one species does show some parental care, foreshadowing the extreme care taken over their offspring by the social wasps. This is *Ammophila pubescens*. The female, as soon as she has mated, constructs a cell in the form of a burrow in sandy soil, extending downwards for about 1in before bending at right-angles to end in a small oval chamber. She then goes off to catch a caterpillar, which she paralyses and drags into the cell and places in the end chamber. She then lays a single egg on it, climbs out and closes the entrance with a sand grain. Then she goes off to construct another similar nest, so that after a time she will have a number to look after. Each one is visited once every day, and after the larvae have hatched and eaten their first caterpillar they need a second, which the female brings them. And so she continues to be occupied until all her larvae have pupated.

Let us end this chapter by considering a special kind of parasitism practised by certain types of bee and usually referred to as social parasitism. The general pattern for this type of parasitism is for the female parasitic bees to lay their eggs in the cells which the host species has prepared for the reception of its own eggs. When the parasite eggs hatch, the larvae are fed by the host workers. Since their egg-laying activities are similar to those of the cuckoo, these parasitic bees are usually referred to as cuckoo bees.

There are two groups of parasitic bees—parasitic bumble-bees which lay their eggs in the nests of other bumble-bees; and parasitic solitary bees, which choose as their hosts other kinds of solitary bees. Britain has two genera of bumble-bees—*Bombus* species, which fend for themselves; and *Psithyrus,* which parasitise them. Like the hive bee, the *Bombus* bumble-bees live in colonies consisting of a single egg-laying female known as the queen; workers who construct the cells in which she lays her eggs and in which the young bees develop, and who also collect the food needed by the queen and the developing young; and males whose only useful function is to fertilise young queens when these are produced. *Psithyrus* species, however, produce no workers, all the work of bringing up their young being under-taken by the *Bombus* workers of the nest in which the *Psithyrus* queen lays her eggs.

The two kinds of bumble-bee are very similar in appearance, but there are some important differences. *Psithyrus* queens, for example, have no pollen baskets and so are incapable of found-ing colonies themselves, since pollen is an essential ingredient in the food of developing bee larvae, providing them with the protein they require. The cuticle, too, of *Psithyrus* is much thicker, making it difficult for the host bees to sting them when they enter their nests. Their jaws are also better developed, and their sting longer and more powerful. The name *Psithyrus,* incidentally, comes from a Greek word which means whispering, and refers to the much softer hum that they make when flying than do *Bombus* bumble-bees.

Like *Bombus* species, only fertilised queens survive through the winter in hibernation, from which they emerge some weeks after *Bombus* queens. These have now founded their colonies and the first few workers have already emerged, so that by the time the *Psithyrus* queens are looking for a nest in which they can lay their eggs there are not too many workers of the host species available to challenge them. Although all *Bombus* species may be generally similar in appearance, there are differences between them. There are similar differences between the various species of *Psithyrus,* and each species of cuckoo bumble-bees chooses the nest of the host species which it most resembles.

Entry to the host nest represents the extreme hazard for the cuckoo queen. She may well be challenged by the workers if she has chosen a nest in which too many have already emerged, in which case she will probably be killed.

The solitary bee cuckoo parasites lead similar lives to *Psithyrus*, and they have the same structural modifications, which make it impossible for them to found colonies of their own. In Britain there are three genera of cuckoo solitary bees—*Melecta*, *Sphecodes* and *Nomada*—and like the cuckoo bumble-bees, they choose host species which most resemble themselves.

CHAPTER FOURTEEN

MITES AND TICKS

The members of the subphylum Arachnida, like the insects, are mainly terrestrial animals, with a small number of species adapted in varying degrees to an aquatic existence. The subphylum contains three orders—the spiders, the scorpions and the mites and ticks. No members of the first two orders have adopted a parasitic mode of life, but the mites and ticks, which together constitute the order Acarina, include many parasites. Indeed all ticks are parasitic at some stage of their life history, but among the mites there are many free-living species. There are some important differences between the members of the two groups, so we shall deal with them separately.

The body of a mite is more or less oval, and it is not possible to identify a head, thorax and abdomen, or a cephalothorax and abdomen, as it is in other arthropods. As in spiders, there are two pairs of mouthparts, comprising a pair of jaws or chelicerae, which usually end in pincers but may be modified for sucking up liquids, and a pair of sensory pedipalps. Although there is no true head, the front part of the body to which these mouthparts are attached is distinguished as the capitulum. Four pairs of jointed legs come further back.

Mite life history involves a number of stages. The larva which hatches from the egg has only three pairs of legs, and in other ways may be quite unlike the adult in appearance. Some species are oviparous, the eggs being retained for development within the oviduct of the female, so that the larvae are born after the eggs have hatched.

In most species the larvae indulge in one good feed before moulting to produce a nymph which has the full complement of four pairs of legs. In some species the larvae moult without first feeding. Most parasitic mites have more than one nymph stage. The first, or protonymph, has one blood meal and moults to produce the second or deutonymph. The last nymphal stage moults again to produce the adult, after which there are no further moults.

Some parasitic mites are just straightforward parasites, feeding upon their hosts and perhaps causing great inconvenience, but seldom doing serious damage or causing death. Others, however, are more sinister, for in addition to their own activities they also act as transmitters of other disease organisms.

Among the most upleasant of the straight parasites are the itch mites *Sarcoptes scabei,* which cause scabies or 'itch' in man. They are extremely small, scarcely visible to the naked eye, but their power to cause irritation is indeed impressive. Itch mites live in the outer layers of the skin, where the fertilised females excavate burrows up to an inch in length which are sunk at an angle to the skin surface rather than vertically. As she excavates the burrow, so the female mite lays eggs at the rate of two or three a day until she has produced a total of between 30 and 50. After these exertions she usually dies at the inner end of her burrow.

After a few days the eggs hatch to nymphs, and these moult in another few days to produce protonymphs, which proceed to excavate burrows for themselves branching off the parent burrow, in the meantime moulting again to produce adults. The whole life cycle takes not more than a couple of weeks, the adults then being capable of living for another 4 weeks or so.

Research has shown that the intense itching is not caused by the mechanical presence of the mites and their burrows. Someone infected for the first time will experience no symptoms for the first month or so. What happens is that the skin gradually becomes sensitised to the mites, presumably through substances they secrete into the burrows. Once this sensitisation has occurred, the irritation becomes steadily worse, so that after about 3 months, if the infection has not been cleared up, the itching becomes continuous and almost unbearable. If a person who has previously been infected and cured is subsequently reinfected, the irritation begins within a day because the skin has retained the sensitivity acquired through the first infection.

Before the cause of scabies was known it was attributed to 'bad blood'—whatever that meant. Great epidemics used to sweep through armies and populations. The modern decrease in its incidence has been due partly to the knowledge of its cause and partly to improve standards of hygiene. Transfer of the mites from one individual to another occurs mainly by physical contact, usually at night from one bedfellow to another.

Related species of mites are responsible for causing mange or scab in many kinds of mammals and birds. In fact in some cases these infections are believed to be due to physiological varieties of *Sarcoptes scabei* which have become adapted to living on other hosts. Once so adapted, it is unlikely that they can then survive if they should find themselves back on their original host.

From the human point of view the mites which causes the greatest harm are those variously known as harvest mites, red-bugs, chiggers or scrub mites, and other names besides. These are species of the genus *Trombicula,* and are widely distributed in many parts of the world. Not only are they extremely irritating little creatures, but in the Far East they rank as a major pest because they transmit the rickettsias which cause scrub typhus. During World War II scrub typhus caused more trouble among the troops than any other animal-borne disease except malaria.

In these harvest mites only the six-legged larvae are parasitic, the subsequent eight-legged nymphs and adults being free-living. As soon as they succeed in obtaining a foothold on their chosen host species, the larvae make a very shallow insertion into the host's skin with their mouthparts, which are modified for piercing, and pass into this hole a minute amount of saliva. This is a powerful digestive liquid, capable of reducing the skin tissue to a semi-liquid product which the larvae are able to suck in. As the first lot of tissue debris is sucked up, so more saliva is poured into the cavity until eventually quite a deep well is formed.

181

Harvest mites do not feed on blood, as one might suppose from their red colour. As with the itch mites, the intense irritation which the mites cause is due to a violent skin reaction to the parasites' saliva. There are many hundreds of harvest mite species throughout the world, the majority of them showing strict host specificity. Relatively few of them will attack man.

Scrub typhus is one of many typhus-like diseases caused by rickettsias. Since only the larval stage of the harvest mite is parasitic, rickettsias can only be transmitted to a host or acquired from a host by the larvae. They must then persist in the body of the mites right through the subsequent nymph and adult stages, and be transmitted to her eggs by the female. Such transovarial transmission of microscopic disease organisms is mostly found in mites and ticks. Very few insect-borne infections are passed on from generation to generation through the eggs.

Another group of mites which cause considerable discomfort to man are the grain mites. They are not parasites but free-living mites which feed particularly upon all kinds of stored foodstuffs, especially grain, flour, sugar, seeds, dried fruits and so on. Economically they are of course important, because heavily infested stores are rendered useless and have to be destroyed. Although the mites will not attack or feed on man, people who work with infested materials often contract dermatitis, which can develop either after contact with the living mites or with dust containing their dead bodies. It probably represents an allergy or sensitivity. The people most likely to suffer from these mites are those whose jobs bring them into constant contact with the materials mentioned above. Hence we have grocer's itch, baker's itch, miller's itch, barley itch and many others.

Although most mites are terrestrial creatures, one group of water mites is present in considerable numbers in almost any stretch of fresh water. These mites are quite active creatures, able to swim and climb among the water plants while chasing their prey, which includes almost any kind of animal life small enough for them to tackle.

As with the harvest mites, it is the six-legged larvae which are parasitic, the other stages, including the adults, living free in the water. So far as is known, the larvae of all species are parasitic. As soon as they hatch, the active larvae swim about looking for a member of their particular host species. If successful, they bury their heads immediately into their hosts' tissues in order to feed on their blood or tissue fluids. The larval legs have served their purpose, so they drop off. Meanwhile the subsequent nymphs are developing beneath the larval skins, and after a time these emerge with a full complement of four pairs of limbs and swim off to lead a free and active existence for some time. Eventually, however, their activity ceases and they enter into a resting stage while the final adults are formed beneath the nymphal skins, to emerge as soon as they are fully formed.

Although the ticks form only one of the five suborders into which the order Acarina is divided, they differ in several important respects from the mites, so that it is convenient to consider them separately. They are much larger animals than mites, easily detectable with the naked eye, which very few mites are without extremely close scrutiny of their hosts.

Like mites, ticks are not divided into separate body regions, and the adults bear four pairs of legs in the adult stage typical of the majority of the arachnids. They have the same two pairs of mouthparts, chelicerae and pedipalps, but projecting forward in the midline beneath and between these the ticks have an additional structure known as the hypostome, which represents an extension of the capitulum. It is a formidable structure carrying many rows of backwardly directed teeth. It functions as a piercing organ which anchors the tick so firmly in its host's flesh when it is feeding that it is virtually impossible to remove it forcibly without tearing its body away from the capitulum, which, with the mouthparts and the hypostome, are left embedded in the host's skin.

All ticks are parasitic during some part of their life history, which typically includes the same four stages already described for the mites. They are among the most important of all animal groups as transmitters of human and animal diseases. The majority of them are mammal parasites, though there are quite a number which attack birds, and some which may be found

183

on cold-blooded animals. On the whole they are less host-specific than some other kinds of parasite. and many of them tend to attach themselves to small mammals or birds in the larval and nymphal stages, transferring to larger mammals when they become adult. This of course facilitates the transfer of disease organisms from the smaller to the larger animals. Rodents and birds often serve as reservoirs of these organisms.

There are two families of ticks—the Argasidae, which are sometimes known as soft ticks, and the Ixodidae, or hard ticks —and the two kinds differ both in their structure and in their life cycles. The body covering of the argasids is a soft leathery cuticle, but in the ixodids the back of the body has a hard dorsal shield which covers the whole body in the males but only the front part in the females.

One of the most common and widespread of the ixodid ticks is the sheep tick, *Ixodes ricinus,* which is also known as the castor bean tick because a female gorged with blood does look rather like a fat bean. The life history of the sheep tick is typical

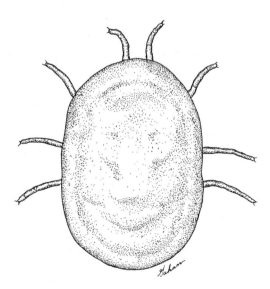

Soft tick *(Ornithodoros moubata)*

of the Ixodidae generally, except that the times elapsing between successive stages of development are much longer than in most other types. Before laying their eggs, the mated and gorged female ticks drop to the ground, where they soon hatch as six-legged larvae that are commonly known as seed ticks.

Further development—indeed ultimate survival—now depends upon finding a suitable host, and the behaviour pattern of these seed ticks is directed towards this end. At this stage they show a strong positive response to light, and this results in their climbing up to the top of grass blades, where they are in a better position to swing themselves on to any sheep which happens to pass close enough for contact. But they do not wait passively. If a sheep moves towards the place where they are waiting,

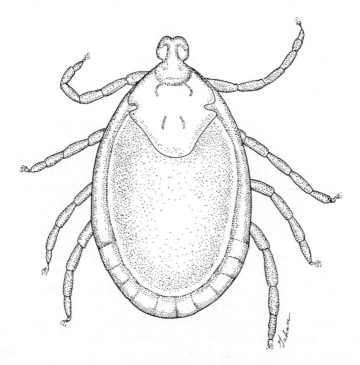

Hard tick *(Dermacentor andersoni)*

185

they become very active. They are believed to be able to detect the approach of a suitable host both through the air vibrations and the smell of the animal. However this may be, they stretch out their bodies to the fullest extent and wave their pair of clawed front legs about in front of them. It only needs a single strand of the sheep's wool to touch one of these questing legs for it to be grabbed. At the same instant the tick's hold on the grass blade by the other legs is released, and what may have been an extremely long period of fasting and waiting is at an end, at least for the time being.

The larva now crawls down the hair until it reaches the sheep's skin, which it penetrates with its hypostome and settles down to its meal of blood. After a few days, by which time its originally rather flattened body will have become round and distended, it releases its hold and drops back to the ground. The instinct to move towards the light, which carried the fasting larva up to the top of the grass blade, is now reversed, and the well fed larva is stimulated to get away from the light among the grass roots.

For some time, which may be for the remainder of the summer and right through the following winter, it remains in hiding, slowly digesting the only meal it will ever take. During this period, beneath the larval skin, an eight-legged nymph is slowly developing. When it does eventually emerge, it is hungry and, like the newly hatched larva, must take steps to find itself a sheep. Its methods of doing so are precisely similar to those already described for the larva. Again, when it has become gorged with its host's blood, it looses its hold and drops back to the ground.

There it remains in hiding while its enormous meal is slowly absorbed and the final adult stage develops beneath its skin. Once again, after a time lapse of anything up to a year, the adult emerges, and again its only hope of survival is to attach itself to a sheep. Of course, so great are the odds of any one larva, nymph or adult being fortunate enough to be able to attach itself to a suitable host, at each stage there will be considerable mortality through starvation of those which do not succeed in finding a host, despite the fact that any stage can go without food for many months following the moult which produced it. For this reason the adult females must lay large numbers of eggs.

It is said that the first concern of the adult female on achieving a place on a host is to search for a mate. Only after she has successfully mated will she settle down to the important business of gorging herself with her host's blood. This, the last of the only three meals *Ixodes* species take during the whole of their lives, is particularly important, because it will supply the materials needed by the female to produce her eggs. Of course, as soon as she has finished the meal, she will drop to the ground, where the eggs will be laid. However many she produces, and some species are credited with being able to produce as many as 18,000, they are all laid at once. The incubation period varies from 2–3 weeks to several months, and eggs which are laid towards the end of the summer may remain dormant until the following spring.

The description just given, based on the life history of the sheep tick, is also typical of Ixodidae generally, but there are variations. Some species have a first nymph stage which, after feeding, moults to give rise to a second stage nymph, and that in its turn produces the adult. In some other species, too, once the larva has succeeded in attaching itself to a suitable host, it does not drop off after feeding but remains while it produces the nymph, so that the latter is spared a hazardous search for its own host. It may in fact also remain on the host to produce the adult. The terms one-host, two-host and three-host are used to indicate the number of times Ixodidae leave their host to find a new one during their lives. Obviously it is a great advantage if the whole life cycle can be encompassed on a single host, because it reduces drastically the losses involved in finding new hosts. It is probably to compensate for these losses that all stages in the life history of a tick can live for long periods. Adults have been kept alive for up to 5 years without ever having a second meal. The part played by certain birds in picking off the numerous ixodid ticks which infest many of the larger game animals has been described in Chapter 1.

The habits of the argasid ticks are different from those of the ixodids. They do not generally live on their animal hosts, or live or lay their eggs in the outside world, but inhabit the homes or nests of their hosts, spending a short time on them when they need a meal. The larvae may make do with one enormous meal, but the nymphs and adults feed frequently, taking in on each occasion more modest amounts of blood. Because they have more or less eliminated the hazards which the ixodid ticks have to contend with, it is only necessary for the females to lay relatively small numbers of eggs, because the resulting offspring are going to spend the whole of their lives in the shelter of the host's home. They do not usually lay all their eggs in one batch, but in a number of batches at intervals of a few weeks.

In Chapter 10 we listed the five groups of micro-organisms which contained parasitic members capable of causing disease. The importance of ticks as vectors of these diseases can be judged from the fact that they transmit members of all these groups. Among the protozoan parasites, as we have already seen in Chapter 6, they are responsible for the transmission of certain Sporozoa belonging to the genus *Babesia*, the causative agents of redwater fever.

A large number of rickettsias transmitted by ticks cause a variety of diseases both in man and animals. These tick-borne rickettsial diseases are sometimes collectively known as tick typhus. A typical example is spotted fever, caused by different but closely related rickettsias in different parts of the world. The spotted fever of North and South America is caused by *Rickettsia rickettsii*, whereas the similar disease known as tick typhus or tick-bite fever in Africa and central and southern Asia is caused by *Rickettsia conorii*. A third relative, found in Queensland, Australia, where the disease it causes is known as tick typhus, is *Rickettsia australis*.

The tick species responsible for transmitting these three species of *Rickettsia* vary from place to place, but they all have similar life histories, which take $2-2\frac{1}{2}$ years to complete. The larvae which hatch from the eggs attach themselves to rodents or other small mammals, dropping off after a single good meal to change to nymphs. These then rest through the following winter, becoming active again in the spring, when they in their turn also seek out a small mammal host. After feeding, they drop off and later moult to produce adults, which also rest right through the winter. When they become active again the next spring, they show no interest in the small mammals which served as their first two hosts, but seek out large mammals, including cattle, deer, horses, dogs and man. The disease organisms are collected from the small mammals either in the larval or nymphal stage, and are then transmitted to the larger mammals by the adults, the small mammals thus serving as reservoirs in which cultures of the rickettsial organisms are maintained.

Incidentally, these rickettsias and the other parasitic microorganisms which we have yet to deal with are not hyperparasites, because so far as we know they do not cause any harm to the ticks, which merely give them a temporary home and pass them from one type of mammal to another.

Two other tick-borne rickettsial diseases worth mentioning are Q fever and heartwater fever. Q fever, an influenza-type disease, was only discovered in Australia in 1937 as a cattle disease, but has subsequently been shown to occur in many other parts of the world. In cattle and horses it is a mild disease, but it is more severe in sheep, goats and man, in which fever with chest pains and cough develops. Heartwater fever is a severe disease of ruminants in South Africa. It owes its name to the fact that an infection results in the accumulation of large quantities of fluid around the heart.

Among the Spirochaetes members of the genus *Borrelia* are the only pathogenic species to be transmitted by ticks. They are the causative agents of relapsing fever in birds and mammals, but only in man is the disease really serious. So far as the ticks are concerned, the various strains and species of *Borrelia* show very strict host-specificity, each species of tick being capable of harbouring and transmitting only its own specific strain. In fact very closely related species of ticks which are so much alike that they are difficult to distinguish between are sometimes identified by the strains of spirochaetes they harbour. Most, if not all, of these tick-borne spirochaetes are transmitted by the ticks via the ovary and the eggs, so they can persist in the ticks for generation after generation without needing to be transmitted to mammals or birds.

Only a few bacterial diseases are known to be normally transmitted by ticks, the most important being *Pasteurella tularensis,* which is the cause of tularemia or rabbit fever. Although primarily a disease of rabbits and rodents, it can also be transmitted to man and to many other animals. The disease symptoms are fever, ulcers and enlarged and painful lymph glands, but the condition is rarely fatal. Perhaps the chief interest is that the bacteria causing it are closely related to the deadly *Pasteurella pestis,* the cause of plague. Ticks have also been shown to be able to transmit undulant fever or brucellosis from cattle to man.

In addition to the harm they cause by transmitting the microorganisms responsible for many different diseases, the ticks are also capable of causing harm through their bites. The most direct effect is severe anaemia if too many ticks are sucking the host's blood. As one might expect, this effect is more likely to be felt by small rather than large animals, but the effects have been observed not only in rabbits and other small mammals but also in deer, horses and sheep. It has been shown experimentally that sixty to eighty ticks on an adult rabbit can impose such a severe drain on its blood that it will die in 5–7 days.

If attempts are made to remove ticks forcibly when they are attached to their hosts, the hypostome becomes so firmly embedded in the skin that in all probability the body of the tick will be severed from its front end, which, remaining in the flesh to decay, may cause sores or ulcers, and even blood poisoning.

A more serious effect of tick infestation is tick paralysis. In both mammals and man a number of actively feeding female ticks attached to the neck or the base of the skull can induce a serious paralysis. There is no evidence that this paralysis is caused by any micro-organisms transmitted by the ticks, but at the same time we have no explanation of how it is caused. It is, however, known that the tick eggs do contain a very toxic substance which might perhaps be contained within the female body and be transmitted in her saliva. Another puzzle is that one individual with ticks attached to its neck will suffer from tick paralysis, whereas another with a similar number of neck ticks will show no sign of the condition. It has been suggested that in order to effect the paralysis one or more of the ticks has to penetrate a nerve ending.

The paralysis usually first manifests itself in the legs, but then travels up the body until it reaches the chest and throat region, which takes 2–3 days. If the ticks are removed, and assuming the heart and respiratory mechanisms have not been affected, recovery takes 6–8 days. Failure to remove the ticks may result in death. Tick paralysis is known to occur both in man and in his domestic animals. Humans affected are usually children, with more girl than boy sufferers, because their longer hair makes it easier for the ticks attached to the neck to go undetected.

CHAPTER FIFTEEN

CRUSTACEAN
AND MOLLUSCAN PARASITES

The phylum Arthropoda generally consists of terrestrial animals which dominate the invertebrate scene on land; but one sub-phylum, the Crustacea, is almost exclusively aquatic, its members dominating the invertebrate kingdom in both fresh and sea water. There are six classes, and only the two most primitive of these have failed to produce at least a few members which have become adapted to a parasitic existence.

The third class, the Copepoda, is a very important group of small planktonic Crustacea which provide a major source of food for many of the larger freshwater and marine animals. Perhaps the most important of them all is *Calanus finmarchicus,* which forms the principal item in the diet of the vast shoals of herring living on the continental shelf surrounding the shores of north-west Europe. The class also contains a number of parasites, some of which have become very modified in adaptation to their parasitic mode of life.

The life history of most Crustacea involves a series of larval stages before the adult stage is reached. The fundamental larva which hatches from the egg is a nauplius, with an oval body and three pairs of appendages—antennules, antennae and mandibles. As these larvae grow and moult, so at each moult additions are made to the hind end of the body. The later larvae in those crustaceans which have a whole series are known as metanauplii, and these in turn may be succeeded by yet other types before at last the adult stage is reached. In the copepods these final larvae—and there may be a series of them, each

192

preceded by a moult—are known as copepodid larvae.

The parasitic copepods show varying degrees of degeneration in adaption to their special modes of life. Some of them infect fish, while others attack invertebrates. Least modified of them all is *Caligus*, an ectoparasite which lives mainly in the gill chambers of fishes, sucking blood from the gill filaments with the aid of a specially modified suctorial mouth. Unlike many parasites, it does retain freedom of movement, and is capable of leaving one host and swimming about in the sea until it succeeds in locating another.

In contrast, *Chondracanthus* shows considerable modification. It is also a gill parasite, but it is so degenerate that it cannot leave its host. In the adult female the appendages are reduced to mere rudiments, and the body is represented by a mass of flesh bearing a number of paired lobes. The males are much smaller and less degenerate, but they remain permanently attached to the females by their antennae. In one species the dwarf neotenaus males are only 1/12,000 the size of the females.

Lernaea is another gill parasite in which the female exhibits extreme degeneration, though only after a rather complex life history. The nauplii which hatch from the eggs continue normal development until the first copepodid stage. These larvae develop a suctorial mouth and become parasitic on the gills of various flatfish, the specially adapted mouth enabling them to obtain food from their hosts' tissues. At the same time the body degenerates, losing most of its crustacean characteristics as well as its ability to move. This phase in its development has been referred to as a pupal stage. It is not, however, the final parasitic stage, for further development results in the production of adult copepods which have fully regained their power of movement. These young adults now mate, and the males do not undergo any further development, having come to the end of their lives.

After this mating the eggs are not fertilised and laid. The females store the spermatozoa for use in the future, while they seek out members of the cod family and attach themselves to the gill filaments of their hosts. This attachment is followed by a major degeneration of the body. The head of the parasite sends processes deep into the tissues of the host, their purpose being to absorb food materials from the host's body. Only remnants of the female parasite's appendages are retained, the remainder of the body developing into a twisted worm-like sac in which large numbers of fertilised eggs are produced.

With most parasites the adults are parasitic and the larval stages free-living, but with *Monstrilla* the opposite is true. The nauplius larvae enter the bodies of various marine worms, their antennae being modified for absorbing food materials from their hosts. The whole of their larval life is completed within the hosts' bodies, but when they eventually change to adults, the latter work their way out of the hosts to live a free life.

The members of the fourth crustacean class, the Branchiura or carp-lice, are closely related to the Copepoda. Until recently in fact they were classed as one of the copepod orders, but the differences between them and the rest of the copepods make it advisable to classify them separately. They are all temporary parasites on fish, where they may be found attached to almost any part of the body. In structure they are perfectly adapted to their parasitic mode of life. A few species are marine but the majority are freshwater creatures, most of them being species of the genus *Argulus*.

Unlike the copepods they possess well developed compound eyes. The mouth is specially modified for sucking blood, and the maxillules bear a pair of powerful suckers with which they can attach themselves to their hosts. The whole body is extremely flattened so as to offer the minimum of resistance as the host swims rapidly through the water. Only four pairs of thoracic appendages are represented, but these are well developed for swimming. The abdomen is much reduced, unsegmented and without limbs. The fertilised eggs are not retained in an egg sac but deposited on stones, where they remain until they hatch. There are no larval stages, the eggs producing tiny creatures essentially similar to the adults in structure.

The most common British species is *Argulus foliaceus*, in which the larger females are about $\frac{1}{4}$in long. This and other species are found on almost every species of freshwater fish, and they are much more common than is generally realised.

The fifth crustacean class comprises the barnacles, and contains one order consisting entirely of parasites known as sac barnacles in which the adult stage is so degenerate that they bear no resemblance either to Crustacea or to any other group of animals. The best known member of the group is *Sacculina carcini,* which parasitises the common green shore crab, *Carcinus maenas.* The external evidence of infection is a yellowish structureless sac wedged between the apron and the shell, forcing the apron away from its usual position in close contact with the shell.

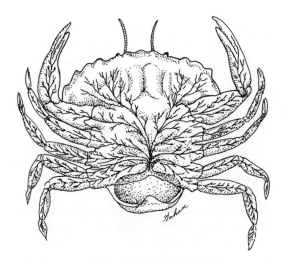

Crab parasitised by *Sacculina*

The appearance of this external sac is the last stage in the life of the parasite, the actual infection having occurred some time before. It contains the fertilised eggs, which are passed out into the water through a small opening at the free end to hatch as typical cirripede nauplii. These nauplii, however, are without mouth or gut, and eventually develop into cypris larvae, which settle on the shore crabs. Each cypris larva attaches itself to the base of one of the crab's hairs or setae by means of one of its antennules. The whole trunk region is now shed, leaving only the head region attached to the crab. From the attached antennule a dart-like structure is produced, and this penetrates the crab's cuticle where it is very thin at the base of the seta. Through the opening thus made the remainder of the larval body squeezes itself into the body of the crab. By now it has become a structureless mass of tissue which is carried round in the crab's bloodstream until it reaches the region of the crab's intestine. Here it attaches itself to the underside of the intestinal wall.

Root-like processes now grow out from the parasite to penetrate every part of the crab's body, even the extremities of its limbs. The food materials needed for this extensive growth are absorbed from the crab's blood. At the hind end of the crab's thorax the parasite develops a sac-like structure consisting essentially of its reproductive organs. This presses upon the ventral integument of the crab on the underside just at the junction between the thorax and the abdomen or apron. Unlike the lobsters and prawns, which have a well developed muscular abdomen that they use in swimming, the crabs' abdomen is reduced to a mere remnant carried permanently tucked beneath the carapace, and is known as the apron. When the crab next moults, the sac pushes its way out of its host's body to become the external egg sac we have already noted, and the life cycle of the sac barnacle is ready to begin all over again. From the initial attachment of the cypris larva to the appearance of the external sac may take the best part of a year. Not only does it take a long time for the barnacle to grow and penetrate all parts of the crab's body, the extrusion of the sac can only take place when the crab moults; and in adult crabs this occurs only once every year.

Generally the host dies, but some interesting effects are produced first. Before a crab moults, it lays down a store of food within its body to tide it over the time when it must hide away from its enemies while its new shell is growing, and to provide the materials for the rapid growth which takes place at this time. Once the parasite has developed to the stage of achieving its external sac, it takes such a heavy toll of the crab's food resources in producing its eggs that the crab can no longer lay down this essential store of food. In consequence it is unable to moult again, and therefore cannot get rid of the acorn barnacles, tube-worms and other sedentary animals which habitually make their homes on crab shells. A crab which has survived a *Sacculina* infection for several years will thus usually be carrying about on its back a veritable living museum of sedentary marine life.

Besides inhibiting moulting, *Sacculina* affects the whole physiology of the crab. The reproductive organs are destroyed so that it can no longer reproduce, a condition known as parasitic castration, which in turn leads to other abnormalities. A young infected female becomes adult in appearance more quickly than normal, while a young infected male gradually loses all its male characteristics as it grows, becoming progressively more female in appearance, with smaller claws and a wider apron. The swimmerets on the apron become fringed with setae as in a typical female, where they are normally used for attaching the eggs while they develop.

Although infected crabs usually die eventually as a result of the infection, a few survive until the parasite itself dies. Reproductive organs are then regenerated, but former males sometimes then become hermaphrodite, producing both male and female reproductive products.

An interesting theory has been put forward to explain the effects of *Sacculina* infection. A female crab produces quantities of fatty substances for forming the yolks of her numerous eggs. These substances have a high food value and form an important part of the food absorbed by the parasite. Insufficient remains for egg production, so the crab's ovaries degenerate through disuse. Male crabs, however, normally only produce small quantities of these fatty substances, but the fact that they are still absorbed by the parasite stimulates the male crab into producing them in even greater quantities, thus changing its metabolism from typical male towards typical female, and inducing corresponding structural changes in the same direction.

A very similar sac barnacle parasite called *Peltogaster paguri* is frequently found infecting hermit crabs, the external egg sac hanging down from the crab's virtually structureless abdomen. The only other member of the order that should be mentioned is *Thompsonia,* which parasitises various species of crabs, including hermit crabs, and also prawns. This is even more degenerate in the adult stage than the other two sac barnacles. When well established, it does not produce a single external sac but many minute sacs which project between the joints of the limbs. They seem only to contain eggs, no trace of testes or spermatozoa ever having been found, and it is suggested that these eggs develop parthenogenetically. They hatch out to typical cirripede nauplii, which are followed by equally typical cypris larvae, the latter settling on the host animals.

Among the class Malacostraca, which comprises all the higher Crustacea, and contains a greater number of different types than the five other classes put together, there are few parasites. These occur in only one of the five subclasses into which the class is divided, and in only two of the five orders included in the subclass.

All but one of these malacostracan parasites are in fact contained in the single order Isopoda, the order which comprises the flattened woodlice and shore slaters. They vary from temporary parasites hardly at all modified in structure for their special mode of life to parasites so modified and degenerate in the adult state as to be unrecognisable as Crustacea.

One of the least modified is *Aega,* a fish louse, which is somewhat heavily built but capable of leaving one host and swimming about until it can attach itself to another. Its mouth appendages are modified to pierce the host's flesh and suck up blood. *Gnathia maxillaris* is another fish parasite, but it only lives and feeds on its host during its larval life, the adults being free-swimming. They exhibit a curious sexual dimorphism for which there seems to be no adequate explanation. The males have a very broad head and enormously developed mandibles, while the female has a head of normal size and shape, with vestigial mandibles. These adults are thought not to feed at all, the food required for their complete life cycle being obtained during the parasitic larval stage.

In another group of isopod parasites there is an unusual type of hermaphroditism, in which each individual first passes through a male phase in which its structure is recognisably that of an isopod, and then changes to a female phase accompanied by varying degrees of degeneration. One of these, *Bopyrus squillarum,* lives within the carapace of the prawn *Leander serratus,* where it appears as a greenish swelling. The male stage is less than $\frac{1}{10}$in long, but it has normal appendages and eyes. The succeeding female stage may grow up to $\frac{1}{2}$in long, but is structurally very degenerate. The whole body becomes more or less oval, with eyes and appendages reduced to minute vestiges, though these are still recognisable for what they are.

Hemioniscus balani, which lives as a parasite inside acorn barnacles, is even more degenerate. Again the minute male stage shows segmentation and has seven pairs of appendages, while the female stage is completely devoid of traces of appendages, and is really nothing more than a bag filled with eggs.

Cryptoniscus presents a very similar picture, but is particularly interesting in that it chooses sac barnacles for its hosts, in this way becoming a hyperparasite.

Isopods are flattened from above, but the members of the related order Amphipoda are flattened from side to side. The best known amphipods are the freshwater shrimps and the various kinds of shore or sand hoppers. One member of the order, and an important parasite, is *Cyamus,* the whale louse, whose body is short and wide. As its name suggests, it is a skin parasite of whales.

The phylum Mollusca contains three main types of animal —the class Gastropoda, comprising the terrestrial, freshwater and marine snails, with a single coiled shell; the class Lamelli-branchiata, which includes the mussels, oysters, scallops, clams and other related types that have a bivalve shell consisting of two separate sections or valves joined by a hinge and held together by muscles; and the class Cephalopoda, comprising

199

the eight-armed octopuses and the ten-armed squids and cuttle-fish.

Very few molluscs have taken to a parasitic life, and those that have belong to the first two classes. There is an important difference between the parasitic members of the two classes. The lamellibranchs are parasitic only in their larval stages, whereas the gastropod parasites are free-living in their larval stages, only becoming parasites when they are adult.

One of the commonest types of freshwater mussel are the various species of *Anodonta*. During the summer the female mussels produce thousands of eggs which, instead of passing straight out into the water with the exhalant respiratory current, as happens with marine mussel species, are retained in special brood pouches situated among the gills. The males do not retain their spermatozoa but allow them to pass out into the water, from which they will eventually be drawn in by the females in the inhalant respiratory water current. Once inside the mussel shells, they are able to fertilise the eggs.

The fertilised eggs remain within the brood pouches right through the winter, developing slowly and eventually hatching out during the following spring as special larvae known as glochidia. Each glochidium already looks like a minute bivalve, with a pair of tiny shell valves and a single sticky byssus thread protruding between them. In this state they are at last allowed to leave the protection of the parent shell. For a time they float about in the water, but eventually settle to the bottom. On their way down many of them get caught by their byssus threads and become entangled in water plants.

Further development is impossible as free animals. The only hope of survival for a glochidium is to become attached to a fish, for the next stage in its life history is parasitic. A fish swimming among the plants may brush against the glochidium, which will attach itself to the fish by its byssus and be carried away. Having been fortunate enough to find a host, the glochidium now proceeds to bury itself beneath the skin, usually on the fins or the tail. It is aided in this by two teeth, one on the free apex of each of its shell valves. The tissues of the fish react to the irritation caused by the foreign body by forming a cyst around it to reduce the irritation, and in this cyst the glochidium can continue its development undisturbed and safe from its enemies, obtaining what nourishment it needs from its host's blood.

Of course only a minority of the glochidia are fortunate enough to find a host. The vast majority wait in vain for a few days and then perish, having no means of obtaining food for themselves. The fortunate few remain within their cysts for about 3 months, while a new adult shell develops beneath the larval shell. The little creatures now bore their way out of the protective cysts and fall to the bed of the pond or river as fully fledged young mussels.

Many freshwater fish serve as hosts to these mussel glochidia, but the three-spined stickleback seems particularly prone to their attack. Sometimes there may be several dozen swellings on its fins and tail, each indicating the presence of a developing larva.

By contrast with the lamellibranchs, all the gastropod parasites are free-living in their larval stages, only becoming parasitic as they reach the adult stage. The vast majority of them parasitise various kinds of echinoderms, which group includes the sea-urchins, starfish and sea-cucumbers, and most of them will be found in the last group. These gastropod parasites of the echinoderms provide a series of examples of external parasites hardly modified for their parasitic existence becoming internal parasites so degenerate that only a knowledge of their development could possibly give any clue as to what kind of animals they were.

These gastropod parasites develop a new organ which is not present in any of the free-living forms. Their proboscis is much more elongated in the parasitic than in the free-living forms, so that it can penetrate deeply into the flesh of the hosts. At its base a frill-like curtain of tissue is produced; this is known as the pseudopallium, the new organ mentioned above.

The least modified of all the gastropod parasites is *Mucronalia palimpedis,* a surface parasite of certain tropical starfishes. It lives on the surface of its host, with its proboscis buried in the host's tissue as far as the pseudopallium. It is permanently attached to its host, being unable to detach itself and transfer to any other host.

Megadenus holothuricola, a parasite of certain species of sea-cucumber, is considerably modified, its digestive system being reduced and its body as well as its proboscis being partially embedded in the host. The pseudopallium is much more developed, surrounding its shell and serving to protect it from the host tissues. The cavity between the gastropod and its pseudopallium serves as a brood chamber in which the fertilised eggs develop.

The next stage is represented by *Stilifer linckiae,* which is so deeply embedded in the wall of its starfish host that only the apex of its shell remains visible at the surface in the centre of a small round opening. There is, however, little structural modification.

Gasterosiphon deimatis has almost reached the limit for an ectoparasite. It is deeply embedded in the body cavity of its sea-cucumber host; an extremely long proboscis is attached to one of its host's blood vessels, from which it obtains all its necessary food requirements; and it retains its connection with the exterior by a short canal representing an extension of the pseudopallium, which envelops the body completely.

All the gastropod parasites so far mentioned are external parasites, because, however deeply embedded they may be in the host's body, they entered through the skin and still retain a connection with the outer surface. The larvae also leave via this opening. By contrast, there are internal gastropod parasites which enter the host's body either through the mouth or the anus, and their larvae also pass out of the body via the gut.

Parallel with the progressive series of external parasites already described is a similar series of internal parasites exhibiting varying degrees of structural modification and degeneration. They are all parasites of sea-cucumbers. The series starts with *Entocolax ludwigi,* which is in fact an external parasite whose internal organs have degenerated almost completely, only the reproductive organs and a brood chamber remaining. The larvae leave the host's body through the external connection.

By contrast, the closely related *Entocolax schwanwitschi* is a

202

true internal parasite. Infection is via the mouth, and the adult female lives in the body cavity of its sea-cucumber host, the larva having bored its way through the host's gut wall. It becomes attached to the outer surface of the host's intestine, but remains connected to the cavity of the intestine by a small opening through which the larvae pass when they are released, thus reaching the exterior through the host's anus. There is a drastic reduction of the parasite's internal organs, only the ovary and the oviduct remaining. In the absence of a pseudopallium, which all the internal gastropod parasites lack, the brood chamber is formed from the dilated end of the oviduct. In this brood chamber live dwarf neotenic males consisting of little more than a single testis and a vas deferens or testis duct through which the spermatozoa are shed into the brood chamber, where they fertilise the eggs.

The last stage in general body degeneration is represented by various species of *Enteroxenos* and *Thyonicola*. Only the body wall, ovary and a short oviduct remain, the general body cavity serving as a brood chamber. There is a single dwarf male embedded in the body wall, its vas deferens opening into the body cavity. The female body bears no resemblance to a mollusc, being a long worm-like semi-transparent tube lying in the sea-cucumber host's body cavity, and having no connection either with the exterior or with the host's gut cavity. Thus the fertilised eggs or the subsequent larvae can only be shed into the host's body cavity.

With any other animal this would mean that they were trapped inside its body until it dies, but sea-cucumbers have an unusual ability which, when exercised, results in the release of the parasite eggs or larvae. When threatened with attack, sea-cucumbers are able to evacuate the whole of their internal organs and, as it were, throw them in the face of their pursuers. The startling effect of this action usually enables the sea-cucumber to make good its escape; and any parasite eggs or larvae imprisoned in its body cavity are necessarily released to the outside world, where their free developmental stages can be completed. Incidentally this drastic evacuation of its internal organs has no lasting adverse effect on the sea-cucumber, for it is able to regenerate a complete new set within a few days.

BIBLIOGRAPHY

Baer, Jean G. *Animal Parasites*. Weidenfeld & Nicolson, 1971

Burt, David. *Platyhelminthes and Parasitism*. English Universities Press, 1970

Burton, Maurice. *Animal Partnerships*. Frederick Warne, 1969

Chandler, A. C. and Read, C. P. *Introduction to Parasitology*. Wiley Toppan, 1961

Gotto, R. V. *Marine Animals*. English Universities Press, 1969

Smyth, J. D. *Introduction to Animal Parasitology*. English Universities Press, 1962

Street, Philip. *Animal Partners*. Edmund Ward, 1958

INDEX

Italic numerals refer to illustration pages

206